Survival Guide
for
Physical Chemistry

Michelle Francl

Library of Congress Control Number: 2001093987

ISBN 0–9713134–0–7

Physics Curriculum & Instruction, Inc.
22585 Woodhill Drive
Lakeville, MN 55044 USA
www.physicscurriculum.com

Table of Contents

Acknowledgments

Many thanks to my husband, Victor Donnay, for all his patience and encouragement, and to my children Michael and Christopher, who continually remind me that learning really is fun and I shouldn't spoil it. I'm grateful to my parents, Lois Cullen Miller and Eugene J. Miller, Jr., for never letting me know that girls weren't supposed to do science and for telling me of their adventures in science. I am grateful also to my friend and colleague Lisa Chirlian, who has shared the teaching of physical chemistry with me these many years. And finally, thanks to all the Bryn Mawr College students who have taken physical chemistry — and lived to tell the tale!

Preface

The American Chemical Society sells a bumper sticker that reads, "Honk if you Passed P-Chem" and while I've never actually seen it on a car, it certainly adorns bulletin boards and office doors in chemistry departments around the country. Physical chemistry is considered by many students to be the make-or-break course of the major. I certainly approached p-chem with trepidation, thanks in part to the tales my parents (both chemists) had to tell about their experiences. My mother boasts of climbing in the basement window of the p-chem lab at dawn to finish off an experiment — this in the days when women students had curfews. Forty years later, my father can still recite the physical chemistry problems on his graduate preliminary exams. ("If water falls from a square faucet, what shape will it be when it hits the sink? Give all relevant equations.") Needless to say, by the time I was a junior chemistry major, physical chemistry had taken on mythical, and not very reassuring, proportions. Imagine my surprise when I found I enjoyed it. Imagine my parents' shock when I decided to pursue it as a career!

Part of the joy of physical chemistry for me is how many other parts of the field can be explored under its rubric. Physical chemists can explore organic, inorganic, and biological chemistry. Subjects as far afield as medicine, art, and anthropology are also legitimate areas for the physical chemist to explore. Truly, physical chemistry can be fun! On the other hand, its breadth and its mathematical underpinnings make it a challenge unlike any other chemistry course.

This book is neither a math book nor a chemistry book. It tries instead to place the mathematics necessary for physical chemistry in a usable context. It's one thing to know the definition of a total derivative, quite another to see how it can be used to derive the Maxwell relations in thermochemistry. Here are also collected many of the bits and pieces learned in earlier courses that can be frustrating and time consuming to track down, such as the rules for rounding learned in general chemistry and the product rule for differentiation hidden somewhere in the calculus text. Finally, I offer some advice that I hope will be useful not only for surviving physical chemistry but also for making the most of it.

Most p-chem courses have a two- or three-semester calculus prerequisite. This certainly goes a long way to covering the math used in modern physical chemistry, but to round it out requires a smattering of other courses as well. Differential equations, linear algebra, probability and statistics, and computer programming courses would be a start. Not surprisingly, few students come to p-chem with all these courses complete. Further difficulties ensue when the student discovers that the knowledge gained in those courses doesn't clearly transfer to physical chemistry. One of the major objectives of this book is to provide the student with a ready reference to the mathematical knowledge base assumed in a physical chemistry course and with ways to bridge the gap between this math and its application to the chemistry. The intent is not to replace the math courses, but to bring together in one spot most of the principles that students need to be reminded of in a physical chemistry class. The basics, such as how to integrate $\sin(x)$, are covered, as are more sophisticated topics, including numerical solutions to differential equations and operator algebra. Particular emphasis is placed on techniques useful in physical chemistry that may not have been stressed in the corresponding math course — for example, the use of a table of integrals.

Computers are a key element in the modern physical chemistry course. Their uses range from word processing to data acquisition. Students arrive in p-chem with varying levels of expertise; they range from experienced system administrators and programmers to virtual novices (pun intended). This book aims to cover most of the typical uses in sufficient depth that a student could, using the appropriate manual, accomplish basic tasks. The details of program use vary widely between applications, but knowing that a task can be accomplished gives the student the incentive to figure out how. If you don't know that you *can* set up a spreadsheet to do a repetitive series of calculations, you won't find out *how*. Topics covered include the use of symbolic math packages such as Mathematica and Maple, because these are coming into broad use.

Simply having the necessary mathematical or computer skills is not enough for a student to be successful in p-chem. Students must also be able to solve problems. Problems in physical chemistry tend to be complex and idiosyncratic, unlike the "plug and chug" problems that often make up the bulk of those encountered in general chemistry courses. Many students appear at my office door complaining, "I don't know how to start!" or "I didn't get the answer in the back, but I can't figure out what's wrong!" A major goal of this volume is to give students some more fruitful techniques than sheer panic to deal with these difficulties.

Writing is another common obstacle. Both the style and format expected in a science course differ substantially from those called for in the humanities. Bibliographic materials are also dissimilar, and students are sometimes intimidated. This book addresses problems that students commonly have in writing for physical chemistry — for example, writing abstracts. It also provides a brief guide to the use of scientific library collections, including electronic materials. Even though students view writing skills as necessary for communicating their results, they often miss the boat when it comes to writing for the sake of documenting their lab work. Guidelines for keeping a laboratory notebook are given, including what to do when you write down the mass of the sample on that little slip of weighing paper instead of in your notebook.

The approach taken throughout this book emphasizes mechanics at the expense of theory. Examples of the techniques discussed are provided, along with heavily annotated solutions. The premise is that being able to reproduce an example is the first step to being sure one understands how to apply a technique. There are no problems to assign; this is meant to be a reference, not an additional textbook. The techniques are meant to be applied to current problems, of which most physical chemistry students generally have plenty!

In writing this book, I've aimed to give students a portable and readable compendium of the information needed in physical chemistry. I also hope they have as much fun taking p-chem as I had 20 years ago, and still have teaching it!

Michelle M. Francl
Bryn Mawr, PA
June 2000

How to Use This Book

Chapters 1 and 4 should be read before you get too far into the course. Both contain some strategic advice that can help you get the most from physical chemistry, regardless of whether anything else ever troubles you with the course. Portions of both of these chapters would be useful refreshers when you face exams or complicated lab writeups. Chapter 1 also contains some strategies that may help when solving problems is giving you a headache.

The remaining material is for reference. I've tried to keep the various sections independent so that you are not forced to read a long diatribe on integral calculus when all you need is a refresher on integrating by parts. Cross references are included so that you can find related information more easily. An annotated list of references is provided as well, for those students who want or need to get deeper into the material.

Chapter 1

Lecture

Though chemistry is very much a hands-on discipline, the lecture is the linchpin of the physical chemistry course. You may be able to do the necessary manipulations in the laboratory, but without a solid conceptual underpinning you cannot easily make sense of the results. Unfortunately, one of the least efficient ways to teach (and thus to learn) is the lecture format! There are, however, many strategies you can use to get more out of a lecture.

You should know first of all that physical chemistry is a time consuming class. Expect to spend more hours per week than you would on the average class, and plan to spend that time regularly. In fact, you will probably spend less time overall if you study frequently than if you periodically try to cram.

1–1 Preparing for Class

On the official Bryn Mawr College teaching evaluation, a question asks the student whether she was "prepared for class." One of my colleagues reported that a p-chem student had replied, "Yes, I always bring a pencil and paper." We both chuckled over this, but we also realized that we had not done a good job of communicating to our students what they needed to do to be prepared for lecture. So what would we recommend?

- Read the syllabus. Be sure you know where the class is. Most faculty don't stray too far from the syllabus — if anything we're usually a bit behind. If you don't have a syllabus, use the reading assignments and your best judgment of your professor's pace to figure out what will be covered in the next lecture.

- Read the book!

  - Keep the reading focused. Don't read much beyond what you think you'll need for the next lecture.

  - Use this first reading of the material to get a broad view of the topic. You should gain some passing familiarity with the terms, equations, and concepts presented.

- Don't expect to understand everything. Don't get hung up on a point you don't understand. Mark it and move on. If you read in small chunks, you shouldn't miss too much of the material that follows.

- Read with a pencil (or pen) in hand. Write your questions in the book. I still do this (just not with library books!). If you just can't bring yourself to do this, or if you are planning to re-sell your text, then invest in some sticky notes. If you have to flip back and forth between your text and a notebook in which you collect questions, you've constructed a barrier you don't need.

- Highlight sparingly, if at all. I often see student texts that are just one solid highlight. If highlighting is your style, wait until you have some perspective on what is really critical, or on what you find particularly confusing, before you go at it.

- Skim the notes from the last class. Mark any major questions you have. Check to see whether the lecturer had any questions she or he wanted you to answer for this class.

- OK — bring pencil and paper! If your lecturer is fond of colored chalk, you might want to bring a couple of different colors of ink or pencil.

This type of approach will work best if you make it a habit before each class. Once the reading starts building up on you, it's easy to get behind. The lectures then make less sense to you, and you begin to get less out of each lecture. At this point it becomes tempting just not to go.

1–2 Attending Lecture

You should go — really! A fellow faculty member made students calculate how much they were paying for each lecture. Her point? If you paid that much for concert tickets, you'd go no matter how sick you were![1]

- **Pay attention.** Once you are there in body, be sure your mind is there too. You can enhance your retention of the material if you are an active participant in lecture. Think about what's going on. Can you see how the instructor got from one step to the next? What happened to the minus sign?

- **Volunteer answers.** There is nothing more demoralizing to a lecturer than to ask a question and get stunned silence in return. Take the risk and offer an answer. Although you might be wrong, you can bet that there are other people who are thinking the same thing you are but just aren't brave enough to say it. Trust your lecturer not to humiliate you (most won't) and trust her to put a stop to it if the answers are all way off base. When I ask a question in class, I'm looking for feedback, to see whether the class grasps the concept, or whether I need to bring in some background material that people are lacking. Answers, right or wrong, give me that feedback and improve the lecture for everyone.

---

[1] You'd at least try to sell the tickets. I have to admit that at this point the analogy fails. I don't see a lot of people hawking tickets for my lectures!

- **Take notes.** Writing engages a different portion of your brain from listening. Research suggests that the more approaches you take to material, the more likely you are to retain it. Some professors provide notes, particularly for material that is quite involved, such as long derivations. You can still take notes. Either ignore the notes handed out and use them later to help fill in your own, or add additional information to the handout. By now you have surely figured out what works for you in the way of notes, but I'll still offer some advice: leave extra space. This allows you to annotate your notes later.

- **Ask questions.** If you're confused about a symbol, or if you don't see where the equation came from, ask! You probably aren't the only one who is puzzled. Try not to preface your question with "I know this is a stupid question...." Some research suggests that women in particular are inclined to structure their questions in this way, but regardless of your gender, don't. If you *know* it's a stupid question, don't ask it! Generally, the only stupid question is "Is this going to be on the exam?" If you think the professor has made an error, let him know. Most aren't offended; I'm usually pleased someone is following closely enough to catch it. (Use your judgment. If the missing minus sign is in a derivation two blackboards back, don't bother!) Take a hint. If the professor puts off your question or suggests that you ask her privately, do that. It may be that the issue will be addressed later in the lecture, or perhaps the question is somewhat off the subject or would take too long to answer. If the lecture gets too chopped up, or if there are too many asides, it can become difficult to follow. That said, follow up on it after class or during office hours.

What if you can't go to lecture? Prepare as though you were going to go, and then get the notes and any handouts from a classmate. Don't expect your professor to provide her notes for you or to save papers for you. In some larger institutions, notes and/or videotapes of the lecture may be available in the library.

1–3 After Class

Go have coffee with your classmates. I'm not joking. Working with a group is another effective technique for enhancing your learning. Some professors will set up groups for you, but the ones you set up yourself can be even more effective. When I took p-chem, a group of us met regularly to work on class material, including one student who was employed as a paramedic. One weekend he participated by phone from a major brush fire — we could hear the helicopters in the background. So go get that coffee, and while you're there...

Annotate your notes. Fill in the blanks in your notes on the basis of your fellow group members' notes. What one missed, another will have gotten. Try to rework any example problems from class. Complain about the long lab writeup due next week.

Later, read the book again. Once again, read with a writing implement in hand. See whether you can answer some of the questions you had on the first reading. Annotate your class notes, filling in additional relevant material from your text. Add the appropriate text page number to each equation in your notes. Work all the example problems given in the text. Many texts include self-test problems that parallel the examples. Try these as well. Finally, check your problem set, and work any problems appropriate to the section.

Take to the next gathering of your group any questions you've generated from your notes or the text and any problems you've had difficulty working. In all likelihood, someone will be able to help you (and you'll be able to help others with something).

1–4 Working in Groups

There is strong evidence that working in groups is beneficial to students, and more faculty at all levels are formally incorporating group work into their courses. Whether you are working informally with a group of friends or formally with a group selected by the professor, there are some common management issues.

A well-oiled group should be interdependent. If one person always provides the answers, the less-active participants won't reap many of the benefits. Faculty generally try to balance the skills of the members in groups they construct, and when you set up a group to work on problems, you can keep this in mind, too. Even if your best friend is a p-chem whiz, don't let her tell the rest of you all the answers. Be sure everyone is clear on the ground rules and on their responsibilities. If you're meeting to discuss the problem set, be sure you've at least looked at it. If you are supposed to have written a section for a group lab report, have it done.

Meet regularly and often. It's easier to keep a standing appointment than to try to get everyone's schedule meshed each time. Meet often so that the work doesn't build up. Try not to meet for the first time the night before the problem set is due. If you've got questions at the end, you won't have a chance to resolve them either within the group or with the professor or teaching assistant. The same thing applies to group projects: Organize early, keep in touch throughout.

Despite your professor's best efforts, you can end up with a dud in your group. If simply telling the offending party to get with it doesn't work (and it very well may not), alert your professor as soon as possible to the situation. He probably can't make the person contribute any more than you can, but he can at least ensure that you don't pay for it. Again, the key here is to intervene early. You're unlikely to get a lot of sympathy if you wait until the lab report is due to say that Kewpie didn't do his part. It's your responsibility to get enough regular feedback to suspect that Kewpie is going to stand you up![2]

1–5 Solving Problems

You've read the book, you've listened to lectures — now comes the moment of truth: the problem set. Although they are connected, knowledge and the ability to manip-

---

[2] If you can do this, you can point out on your résumé when you graduate that you have excellent management skills!

ulate and apply it are not the same thing. When my cats were kittens, they were caught in a house fire. The only damage done was a coat of ash, but because cats clean themselves by licking, I thought they should have a bath. My cat-care book, however, said that kittens should not be bathed; they could lose too much body heat. A quick call to the vet yielded the following instructions: In a warm, draft-free room, run a sink full of warm water, calmly immerse the kitten, pour water over it, soap, rinse, and dry in a warmed towel. The vet then laughed and said, "Good luck!" I had the knowledge, but applying it was another story. By the time I was done I was covered in scratches. On the positive side, subsequent baths haven't been nearly so awful. Encounters with physical chemistry problem sets are not so different. The early ones can leave you a bit bloody, but after a while you learn how to duck the claws.

Realizing three things can help make problem solving of almost any sort go better. First, problem solving is a skill, and as for most skills, practice will help. Next, there is more than one way to proceed in the solution of almost any problem. Finally, a little knowledge is a dangerous thing.

As with cat baths, regular repetition of problem-solving skills hones them more precisely and more rapidly than occasional practice. Practice means more than just working the assigned problem set. It includes taking the time to revisit problems you had difficulty with in the past. I highly recommend setting aside time to work on problems at least three times a week. Larger blocks of time are generally better than smaller ones, and it is easier to remember what you were doing if a long period has not elapsed since you last tackled the darn thing. This all requires much less effort on your part if your instructor collects your problem sets. Many do not, however, and in these cases it's up to you to be disciplined enough to do the assigned work in a timely fashion.

There is a perception among non-science students that all science and math questions have one, and only one, correct answer and that it's usually a number. It's helpful to realize that this is actually true only of simple problems that are very well defined. Many more problems have a variety of solutions. The final answers obtained may even vary depending on the assumptions one makes. The moral: If one method isn't working out for you, try another. For example, trying to get my two-year-old to open his mouth so we could brush his teeth required several attempts at a solution. The need to keep his teeth from getting "boo-boos" didn't move him, but cleaning the frog in there, now that works! (A college education can be useful in ways you never imagined!)

Finally, don't try to solve problems in a vacuum. In other words, read the book, listen to the lectures, and have some background to support your efforts. Recently, a sailor in a solo around-the-world race had a badly infected arm that needed to be drained. The nearest emergency room being hundreds of miles away over open ocean, a physician gave him long-distance instructions for some do-it-yourself surgery, which went amazingly well. Later the sailor noticed a lot of bleeding, and wrapped a cord around the arm to staunch it. Unfortunately, he left this tourniquet on too long and ended up with some nerve damage. A little knowledge may not get you all the way through!

All this advice sounds very "teacherly," but I learned these lessons the hard way — as a student. My freshman year I took intro physics, where problem sets were collected but didn't count for very much. I did the problems I could and skipped the rest. I never looked at the problem sets again until the final, where I did try the hard ones, again without much success. Needless to say, the final did not go particularly well, and my grade in the course was a humiliating F. Welcome to college! Although I didn't learn a lot of physics that term, I did learn some other useful lessons. When I repeated the course, I was extremely disciplined about the problem sets and managed an A.

Problem-Solving Techniques

Enough nagging — time for some practical advice! Here is one approach to tackling recalcitrant problems:

- **Write.** Staring at a blank piece of paper is demoralizing. Even if you just write down the questions you have or sketch pictures, writing will help to focus your thinking.

- **Dissect.** What is the essential question or questions being asked? What result is desired? A lot of extraneous information is often collected in a problem. Winnowing the question from all the chaff helps get you focused.

- **Define.** Are you sure you understand all the symbols and terms used? What units and conditions are specified in the problem? You may need to hunt for clues to these. For example, if you have a mixture of ice and water in a beaker on the table, it's nearly a sure bet that the temperature is 0°C, even if it isn't directly specified.

- **Draw.** Sketch the experimental setup described. Make a schematic diagram of the thermodynamic cycle being described. Plot the potential energy function specified.

- **Outline.** Just as when you write a paper you put together an outline of the work, you can sketch a solution to the problem. Break down the question into several intermediate steps, if possible. If you can't see where to go, try to work backwards. Think of a way to get the result you need, even if you don't think you have the necessary information to get it. Be sure you are specific about what you are trying to do at each step.

- **Data.** What data is given in the problem? What should you make assumptions about? Remember that some information may not be explicitly stated. You may need to make reasonable assumptions. Are there useful tables in your book or otherwise accessible to you? In physical chemistry, as in real life, not all the data you need is contained in the problem.

- **Do it.** Just try it. See how far you get. Do you need more information? Have you run into a snag mathematically? Annotate your solution. Describe, in words, what you are doing and why. This helps to reinforce your strategy (multiple modalities increase learning!). The remarks will become essential when you try to go back to the problem later or as you try to figure out where you went astray!

- **Refine.** If need be, refine your strategy. Is there a way to get around the point where you're stuck?

- **Decide.** Is the answer reasonable? Should it be positive, negative, real, complex? What is the rough order of magnitude? Are the units correct? This step may seem trivial. After all, you can often just look it up in the back of the book. However, because real life doesn't come with a back of the book, this skill is as important as coming up with the answer in the first place![3]

All this is easier to see than to describe. Following is an example of a real physical chemistry problem whose solution I've approached using the method described above.

> The isothermal compressibility of lead is 2.21×10^{-6} atm^{-1}. A cube of lead having sides of 10 cm at 25°C is to be inserted in the keel of an underwater exploration TV camera, and its designers need to know the stresses in the equipment. Calculate the change of volume of the cube at a depth of 1.000 km (disregarding the effect of temperature). Take the mean density of sea water as 1.03 g cm^{-3}. Given that the expansion coefficient of lead is 8.61×10^{-5} K^{-1}, calculate the volume of the block, taking the temperature into account too.[4]

- **Write.** Grab that big piece of paper. Start writing — anything! "Why should you want to put a cube of lead on a camera? That stuff is heavy!" "To make it sink... that's the idea of an underwater camera — it sinks!" "Got it...." Anything is better than staring at a blank sheet of paper!

- **Dissect.** What do they want to know? Two things: (1) The change in volume of a lead block at a constant temperature when it's taken down to a depth of 1.000 km underwater. (2) The total volume of the block under the same conditions, but this time include the change in temperature.

- **Define.** Isothermal — hmm. Translates as "constant temperature". Isothermal compressibility? Expansion coefficient? Ah — useful formulas in the chapter:

$$\alpha = \frac{1}{V}\left(\frac{\partial V}{\partial T}\right)_p \quad \text{(expansion coefficient)}$$

$$\kappa_T = -\frac{1}{V}\left(\frac{\partial V}{\partial p}\right)_T \quad \text{(isothermal compressibility)}$$

I don't immediately see any use for these things, but I write them down anyway.

- **Draw.**

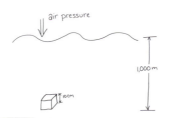

---

[3] My general chemistry students are usually aghast when I tell them that I will refund all the points they would have missed if, when they reach the end of the problem, they realize the answer is incorrect and say why. My point here is the same: Learning to tell when you should have confidence in your solution and when you shouldn't.

[4] P.W. Atkins, *Physical Chemistry*, 6th ed. (New York: Freeman, 1998), page 93.

• **Outline.** OK — I want to know the change in volume. This is p-chem, so it's probably not going to be some simple ΔV thing. No useful formulas come to mind, and I don't see anything worthwhile as I browse through the text again. Let's see if I can find dV and integrate it over the limits.

$$\text{change in } V = \int_{\text{starting conditions}}^{\text{final conditions}} dV$$

That's a big help?? What's dV? V depends on what here? Pressure and temperature! OK, if I've got $V(p,T)$, then I know from the definition of the total derivative (a trick used in the chapter for something else!) that

$$dV = \left(\frac{\partial V}{\partial p}\right)_T dp + \left(\frac{\partial V}{\partial T}\right)_p dT$$

So let's try the following: (1) write dV; (2) find expressions for the partial derivatives; (3) figure out the pressure and temperature conditions over which to integrate; and (4) integrate.

• **Data.** I need to know the temperature and pressure values over which to integrate. I'm going to assume the initial pressure is 1 atm — there's nothing in the problem to suggest that the camera is going to be assembled in Denver. What's the pressure at a depth of 1.000 km? Check the index and find that there is a formula for the hydrostatic pressure:

$$p = \rho g h$$

Not so coincidentally, the problem provides the density of the sea water! I must be on the right track. So that'll give me the pressure due to the sea water. I need to remember to include the pressure of the air above that, but that's easy; it's just 1 atm (again, presuming the camera isn't going to be used to explore some Andean lake!). The temperature changes are given. For (1) I can assume it's 0, so $dT = 0$, simplifying the entire deal.

• **Do it.** First things first — the isothermal case.

$$dV = \left(\frac{\partial V}{\partial p}\right)_T dp + \left(\frac{\partial V}{\partial T}\right)_p dT$$

because $dT = 0$,

$$dV = \left(\frac{\partial V}{\partial p}\right)_T dp$$

Then, using the expression for the isothermal compressibility, which is

$$\kappa_T = -\frac{1}{V}\left(\frac{\partial V}{\partial p}\right)_T$$

I can solve for an expression for the partial derivative I need:

$$\left(\frac{\partial V}{\partial p}\right)_T = -V\kappa_T$$

substituting this into my expression for dV yields

$$dV = -V\kappa_T dp$$

Next, I rearrange to get all volume terms on one side and all pressure terms on the other.

$$\frac{dV}{V} = -\kappa_T dp$$

Now integrate — but hey, I don't know the final volume. Ah — I'll solve for that.

$$\int_{1000cm3}^{Vfinal} \frac{dV}{V} = -\int_{1atm}^{1atm+\rho gh}\kappa_T dp$$

From here, you should be able to see that I'm going to get an expression for V_{final} on one side and something in terms of κ_T on the other and be able to solve for V_{final}. I'm on the alert about units — the given value of κ is in atm^{-1}, but it'll be easier to get the volume if I put everything into SI units, so I'll convert κ_T to Pa^{-1}!

The second half of the problem works similarly, where now you have to integrate over both pressure and temperature.

- **Decide.** The actual numerical answers I get are for the final volumes are (1) 985.8 cm^3 and (2) 985.5 cm^3. These make sense. Lead is pretty incompressible, so the changes from the initial volume of 1000 cm^3 should be small. Increasing the pressure should decrease the volume, and my answer reflects that. Decreasing temperature also generally decreases volume;[5] again, my answer makes sense. Check the number of significant digits and units. All are OK.

While I wrote this, I read Ernest H. Shackleton's account of an Antarctic expedition.[6] He describes his party's frustrations with one of its pieces of scientific apparatus that had been constructed with an aluminum housing (to keep weight down) and brass screws. The screws fit tightly into the body at normal temperatures, but because the expansion coefficients of the two materials are different, the screws were loose at the extremely low temperatures at which they were using the apparatus. If you are planning any polar expeditions, p-chem might actually turn out to be useful!

---

[5] Water is an exception to this; over certain ranges, decreasing temperature increases its volume.

A couple of points: I had to make some assumptions where the problem didn't make things clear. For example, I assumed that the pressure of the atmosphere above the surface of the water was 1 atm. Also, the decision to use the total derivative for dV seems to come from nowhere. However, if you had read the corresponding chapter in the book, you would have noticed another problem that was solved by pulling the total derivative out of a hat. Often the "trick" you need to solve a problem is there in the chapter already — just in a slightly different form.

Solving many of the physical chemistry problems you come across, both now and later in your careers, will require that bit of inspiration. What you need to have are techniques that help you find that creative spark. There really is an art to science!

Essay Questions

Physical chemistry isn't always about crunching numbers. You should also be able to produce written descriptions and explanations of phenomena for a variety of audiences. What do you tell your dad when he sends you an advertisement for an instant food defroster? (It violated the second law of thermodynamics big time!) Can you explain, in a technical way, why it's easier to drink from a straw at the North Pole than on top of Mt. Everest?

Approach these essay questions in much the same way you'd approach an essay question in English or political science. Keep in mind that your aim is to convince the reader succinctly of your point or points. Avoid dumping in everything you know. If I have to look too hard for the answer in pages of discursive prose, I'm not a happy professor! View these questions as a good way to show off what you know without having to work your way through pages of equations.

The Solutions Manual

> To have, or not to have: that is the question:
> Whether 't is nobler in the mind to suffer
> The slings and arrows of a grader's fortune,
> Or to take arms against a sea of troubles
> and crack the solutions manual....

OK, so I'm not the bard! You may not be able to choose whether to have or not to have (the solutions manual, that is), depending on your instructor's preferences. I generally don't order them for the students to buy at the bookstore, but I do put one on reserve for them in the library. Used properly, they're a great resource for learning physical chemistry. They are also a great resource for not learning physical chemistry. Given that you have one to use, what can you do to be sure you're in the former category?

- Put the solutions manual somewhere inaccessible. The easier it is for you to look at it, the more often you are likely to do so. Leave it in your backpack — or better yet, in your room — while you work problems. Play by Scrabble rules, where

---

[6] *South: A Memoir of the Endurance Voyage* (New York: Carroll and Graf, 1998).

you are docked points if you consult the dictionary and are wrong! If you need to, set up a real penalty for yourself: 10 sit-ups, 50 cents, whatever works.

- Never use the manual until you've filled several empty sheets of paper. If you can't even begin working the problem, you need more help than the solutions manual can provide. Try reading the appropriate section in your text and/or getting some help from a fellow classmate. If that doesn't work, it's time for office hours, not for a stroll through the solutions manual.

- Don't read the whole solution. Don't let everyone in the group read the solution. That's like reading a review of a mystery novel that gives away the villain. Try to read just enough to get the clue you need to proceed. If you're working in a group, designate one person to check the solutions manual; then have him or her share the clue with the rest of the group. The person who presents it to the rest gets the material reinforced, and the rest of you don't have to resist the temptation to look further than you should.

- Mark the problems for which you had to use the solutions manual. Rework them later without referring to the solutions manual. If you can't do that, you're missing some key piece. Try to figure out what the fundamental difficulty is and to address it. Is it that you don't remember what adiabatic means? Are you having trouble differentiating something?

- Have confidence in your own approach. Just because you used a different method to solve the problem doesn't necessarily mean you are wrong. Most of the problems assigned in a physical chemistry course can be solved using more than one technique. Finally, every once in a while, the answers in the solution manual turn out to be wrong. No one is perfect — not you, not your professor, not the textbook author.

- Finally — we can tell. I generally know when a student has copied the answers to a problem set from the solutions manual.

Miscellaneous Advice

Some people find it easier to work with a blackboard, even when working solo. If you are the sort who likes to spread out and can't always find a spare room with a blackboard to write on, try artists' large pads of newsprint. They are a bit bulky to carry around, but unlike work on a blackboard, this work can be saved!

Trying to get a group together for a weekly problem session can be daunting enough, to say nothing of arranging for more-frequent meetings. This can be particularly tricky when many students live off campus. Alternatives include "phone time," where you schedule a time for a quick chat about sticky points with a partner. You can try the same thing with e-mail, arranging to check it by a certain time.

1–6 Getting the Most from Your Professor

One of the things I like best about teaching is the interaction with students. It can also be a source of tension. There was the time a student followed me into the ladies room, calling "Dr. Francl, I know you're in there!" Another student wanted to know what hospital I was going to be in for surgery, so she could "call with any questions I have before the [10 point!] quiz." Thankfully, these students are a tiny minority! How can you avoid being part of your professor's collection of worst-student stories?

- Professors are people, too. They have families and pets and need to shop for groceries. The same goes for teaching assistants. Treat them as people and they are likely to treat you the same way.

- Know when the professor's and TA's office hours are. Respect their schedules. Don't assume they are necessarily free to answer questions just because you see them in the hallway or with their door open. If you can't come to office hours, see if you can make an appointment for another time. If you can come to office hours, do. Don't use appointments as a way to get around waiting in line during office hours.

- If you have a question about a course policy, check the syllabus to see whether you can clear it up. If that doesn't work, check with the professor directly. Don't rely on colleagues or even on the TAs to accurately relay course policy.

- Professors do more than just teach your course. Often they are teaching more than one course. They also sit on college committees, do research, apply for grants, write research papers, supervise graduate and undergraduate research students, give outside lectures, and work with professional societies. A student once wrote on my evaluation that every time she came to office hours I was busy. She seemed surprised and put off by this, but I just don't have the time to sit idly waiting for visitors! Your professor, too, will probably be busy when you come to office hours, but he should still be perfectly willing to stop to work with you. Don't be intimidated — knock! If a professor isn't at a regularly scheduled office hour, hang out for a couple of minutes to see whether he returns. He might just have gone for a soda!

- If you need to meet with the professor privately to discuss confidential issues, let him know. Most of us won't make you discuss your grade or plans for graduate school in front of the entire class.

- Tragic and uncontrollable things happen. In 15 years of teaching I've seen students who've had babies during finals, temperatures of 106°F, and other crises of major proportions. Generally speaking, physical chemistry is the last thing on your mind in a crisis, but try to have someone communicate with the professor so that arrangements can be made for you to complete the course. Most professors will work with you under circumstances like these.

- Less-than-earth-shattering things happen, too. If you think you'll need extensions or other accommodations, let your professor know as soon as possible. I'm more likely to give an extension on the end of term project to the student who has been struggling all semester with vision problems than to the student who had the flu the last day of the term. Be prepared to hear "no" and have a backup plan. Every professor has her definition of what constitutes an unforeseeable emergency. Remember, lack of planning on your part doesn't constitute a crisis on my part.

Office Hours

You're in the door — now what?

- Get organized ahead of time, know what your questions are, and have them prioritized in case you can't ask them all. Bring your notebook or problem set with you if that's where your questions lie. Know what page the confusing equation is on in the textbook. Write this stuff down if you have to.

- Try to be as specific as you can about what you need to know. The poor general chemistry student who appeared in my office to say that she just didn't understand "bonding," which we had spent several weeks covering, was hard to help. Office hours aren't intended to replace reading the book or attending lectures.

- Be open to working with a group. I often invite the three or four students waiting to see me to come in as a group. Often they have similar questions, and I can spend a bit more time answering them if I do it for the entire group. Of course, if you need to have a private chat with the instructor, let him or her know!

- Be respectful of other students' time. If there is a long line outside the door, ask only your most critical questions — or better yet, offer to share your time.

Other Access

E-mail is often a good alternative to office hours. Find out what your professor's preferences are. For example, I generally tell students that questions submitted by e-mail before 8 p.m. will be answered before 10 p.m. Other people answer questions on a rolling basis. E-mail can be particularly good for getting procedural questions answered ("What is the deadline for the kinetics lab report?") but don't let that stop you from trying to get more substantial questions answered. Some courses run electronic lists or newsgroups for students so that they see the questions others are asking. Even if you just "lurk" and don't participate, these can be a helpful resource.

1–7 Exams: Strategies and Stresses

This is what it all comes down to — the exams. Exams are generally stressful, Jason Fox in FoxTrot notwithstanding. I took a philosophy course a few years back, just for fun. When the first exam came, I discovered that even though it didn't "count," my palms were sweaty and I was breathing fast. Exams are stressful. A wide variety of instruments can be used to assess what you have learned in physical chemistry, and your professor may use one or more different formats during the semester. Exams can be given in class or out of class, they may have a time limit or not, and you may or may not be able to use your text or perhaps only a sheet of notes. Regardless of the format of an exam, you need to prepare for it. There are also some specific strategies that can help you do your best on different types of tests.

What's on the Exam?

Please don't ask your lecturer, "Is this going to be on the exam?" Would we lecture about it if we didn't think it was important? Do we think you'd listen to the rest of the lecture if we said "no"? Of course not! What can you do, then, to find out what will be on the exam? It's generally OK to ask what the exam will cover: the first six weeks of the term, Chapters 4 through 10, the whole term? Because I can't test everything in an exam, I usually focus on the most-important things. You can generally prioritize topics by the amount of attention paid to them. If a topic was assigned as reading, appeared in lecture, was represented by problems to be worked, AND was needed in a lab, you can almost bet that it will appear on the exam. At the other end of the spectrum, if you saw it only in the reading, the probability that it will form a substantial part of the exam is much lower.

Style

Every professor's tests are a reflection of his style. I love to ask questions that apply to real life and are full of red herrings. Other people in my department ask questions that are more straightforward. Some use a mix of essay questions and numerical problems; others focus strictly on problems. Some faculty put sample exams on reserve in the library. Sometimes you can get information about an instructor's testing style from students who took the course last year.

And sometimes you just have to wait and see! The techniques that follow can help. I've used them in philosophy and theology classes in which I generally didn't have a clue what the instructor's testing style might be.

Preparing

If you've been doing all the things recommended in the rest of this chapter, you shouldn't have to spend an inordinate amount of time studying for mid-term exams or finals. Even if you haven't been doing everything I recommend, the following approach should help.

Outline the major topics to be covered on the exam. If you're very lucky, you're professor will have provided one for you. If you're not one of the fortunate minority, you're still lucky, because making the outline is a terrific way to review. Prioritize the topics in terms of (1) the emphasis your instructor placed on them (see above) and (2) your comfort level with the topic.

• Start early —a week before the exam date, if possible.

- Make a formula sheet, using your outline of major topics. Be sure to note any restrictions that might apply to an equation (such as constant temperature). Remind yourself of the location of any critical data tables (such as the page on which Henry's Law constants can be found).

- Practice with problems related to the topics above under exam conditions (open book, closed book, etc.). Keep style in mind; generally open-book exams have longer, more intricate problems. Focus on your top study priorities: topics that the professor has emphasized and that you are least comfortable with. This last strategy will help you make the most of always-limited preparation time.

Taking an Exam

Several principles apply to taking an exam, open book or not:

- First, read the directions.

- Then do the problems.

- Finally, check your answers.

The first sounds pretty obvious, but sometimes students get so excited (OK, panicked!) about taking an exam that they skip step 1. Be sure you know what you have to do — answer four out of ten questions, plot this, write a paragraph on that. I often ask students to write a line or two about the numerical answer to a problem. I'm amazed by the number of students who lose a couple of points here and there for failing to write that comment. Step 2 sounds even more obvious than the first. All I mean by this is that you should not forget the problem-solving techniques that you generally use and that work for you. If you find working in small spaces on exams confining, bring large pieces of blank scratch paper. Take time to outline the solution. Finally, make sure all your answers are reasonable. Pay attention to orders of magnitude and signs ("Should that pressure really be negative???"). Did you use the appropriate number of significant figures? (This is the most-commonly-made error on all my exams!) Did you specify units when they were required? Are the units correct (not liters for a pressure, for example)? Check to see that any prose answers make sense.

Open-Book Exams

Many students (including me, I admit) heave a big sigh of relief when they learn that an exam is open book and/or open notes. Memorizing "stuff" that I may never use again is not my idea of a good time. I find that if I use it often enough, I remember it, and if not, I just look it up. As a result I frequently use open-book exam formats. The critical point to remember about open-book/open-note formats is that they should not be used as a pretext for failing to study the material the first time. Nor should you forgo any aspect of exam preparation. Here are some useful strategies for taking an open-book exam (in addition to the general guidelines given above):

- Use a summary page to help you find material in your notes and text quickly. The formula sheet you wrote while preparing to take the exam should come in handy at this point.

- Use what you have access to! If you have a table of integrals in the back of the book, don't re-do it from scratch. If a result you need is given in a example problem, cite your source and use it.

- If you use things from the book, be sure to note the page or table number. This makes the grading task easier, and a happier grader is a thing to be desired.

One Page of Notes

My family teases me about my microwriting, but I've seen students who can put me to shame. Up to a point, being able to fit a lot on that precious one page of notes is important, but be sure you're not just creating a microfilm version of the text. Use your prioritized list of topics and the prepared formula sheet to guide you. Be particularly careful to note the conditions under which an equation can be used. I can't tell you how many times I've seen students use an expression for the density of an ideal gas to compute the density of a liquid or a solid. If a formula requires constants, be sure you've got those jotted down somewhere, too. Conversion factors for frequently encountered units are also a useful addition to your page. Under these circumstances, you can't necessarily count on your instructor to provide them.

Exams Without Time Limits

Should you sleep? Yes. Sleep is generally a good thing. A good strategy for an exam with no time limits is not to spend every waking second working feverishly on it. Instead, split your working time up into two or three blocks. The first time through, finish off everything you can do easily. Take a break from the exam — for at least an hour, if you can spare it. Things that didn't seem clear the on the first pass you may find you can do now. If you can manage it, take a second break and then come back and check your answers.

Chapter 2

Guerilla Math

By far, the biggest problem students encounter in physical chemistry is math. This chapter is not intended to replace the two or three semesters of calculus that are typically required for p-chem, nor do I make any pretense of teaching you mathematics. This is guerilla math — it's not elegant, but it should get you through!

The various sections in this chapter are designed to remind you of things you already know (for example, integral calculus) and to add to your mathematical armory some pieces that are useful for p-chem. The material is presented with physical chemistry in mind, and rarely is any attempt made to justify it theoretically. Just believe me! I've also tried to collect various items of trivia, such as conversions between Cartesian and polar coordinates, the formula for the surface area of a sphere, and trigonometric identities, that might prove useful as you work your way through problems. There are also brief summaries of some very basic material — dealing with logarithms, for example. Experience suggests that you tend to forget the simplest things at 1 a.m. in the midst of a problem set. For the curious, and for those whose courses are taking them deeper, the references at the end of the chapter offer more-detailed information on many of the subjects included here.

2–1 Know Your Symbols

Knowing how to execute the appropriate mathematical operation is only a part of solving the problem. Before you begin, be certain you know what the various symbols mean in context. For example, does the Z in the equation refer to the collision number or the compression factor? Is n a vibrational frequency or a stoichiometric number? Does the bar across the top of the symbol refer to a vector or does it mean the average? Different textbooks adopt different conventions; be familiar with yours, and check carefully when referring to other texts. Your instructor may have preferences that are at odds with those found in the text, so be alert to these, too. I try to be consistent with the text I am using, as do most faculty, but old habits die hard!

When writing easily confused symbols, such as w and ω, be sure to distinguish consistently between the two. This is probably nowhere more important than in quantum mechanics when you work with the harmonic oscillator problem. Many a problem has gone down in flames when v (referring to the quantum number) and ν (referring to the vibrational frequency) were confused.

During long car trips when I was a child, my mother amused us by teaching us the Greek alphabet. For those who missed out on this formative childhood experience, here it is!

A	α	alpha	H	η	eta	N	ν	nu	T	τ	tau
B	β	beta	Θ	θ	theta	Ξ	ξ	xi	Y	υ	upsilon
Γ	γ	gamma	I	ι	iota	O	o	omicron	Φ	φ	phi
Δ	δ	delta	K	κ	kappa	Π	π	pi	X	χ	chi
E	ε	epsilon	Λ	λ	lambda	P	π	rho	Ψ	φ	psi
Z	ζ	zeta	M	μ	mu	Σ	σ	sigma	Ω	ω	omega

2-2 Numbers

Real, complex, rational — what's what with numbers?

Integer

...–3, –2, –1, 0, 1, 2, 3.... A more-technical definition can be found by checking some of the references in Section 2–14. Some professors will designate this set by $\mathfrak{J}$, others by **Z**. Most blackboards don't include a "bold" function, so be sure you know what your instructor uses in writing!

Rational and Irrational

Any number that can be expressed as a ratio (hence the name) of two integers is a rational number. Irrational numbers cannot be expressed in this way.

Real

The set of all rational and irrational numbers is the real numbers, which are typically designated by using the symbol $\mathfrak{R}$.

Complex

The set of all numbers of the form $x + iy$, where x and y are real and i is $\sqrt{-1}$, are called the complex numbers. The complex conjugate of a number a (indicated by a^*) is given by

$$a = x + iy$$
$$a^* = x - iy$$
$$aa^* = a^*a = x^2 + y^2$$

Others

There are also perfect, abundant, and deficient numbers, but because they don't show up in introductory p-chem, you'll have to go ask your math professor if you're curious.

2–3 Series

Series crop up in several broad areas in physical chemistry. Products of series appear in quantum chemistry, as do sums of series. Sums of series are also quite common in the formulations used in statistical mechanics. Most of the troubles that students have with sums and products of series stem from misunderstandings about the notation, so if a problem with a summation is giving you headaches, check below to be sure you are clear on what you can and cannot do with it.

Summation of Series and Products of Series

Simple Sums of Series

Basically this is just a compact notation for generating and adding a list of numbers. Compact notation is great when you have to write something over and over again, but it is a pain when you actually have to manipulate the notation and come up with a number. The basic notation for "add up a bunch of numbers" is

$$\sum_{i=start}^{finish}(\quad)$$

which instructs you to begin by substituting *start* for all instances of *i* in the expression within ( ) and evaluating the expression. Repeat after incrementing *i* by 1, and add what you have to the first value. Continue incrementing it and adding subsequent values to the current total. When *i* equals *finish,* evaluate the expression, add its value to the total, and stop. Here *i* is called the summation index. The index is always an integer, as are the limits of *start* and *finish*. The summation may have an infinite number of terms. This procedure is much more easily shown than described, so here it is:

$$\sum_{j=2}^{4}(2j)^2 = (2\cdot2)^2 + (2\cdot3)^2 + (2\cdot4)^2 = 116$$

$$\sum_{i=1}^{5}x_i^2 = x_1^2 + x_2^2 + x_3^2 + x_4^2 + x_5^2$$

Other Forms You May See

One of the most frustrating things about compact notation is that people keep trying to compact it even further. If you are the one writing it ten dozen times, this seems fine, but readers (including yourself at a later date!) may assume you've done it just to torment them. For example,

$$\sum_i (\ )$$

might mean

$$\sum_{i=1}^{N}(\ ) \quad \text{or} \quad \sum_{i=1}^{\infty}(\ ) \quad \text{or even} \quad \sum_{i=0}^{N}(\ ) \quad \text{or} \quad \sum_{i=0}^{\infty}(\ ).$$

The compactor might go one step further and simply write

$$\sum (\ )$$

in place of any of these expressions and leave you to assume that any integer that appears as a subscript or superscript, perhaps designated by the letter i, j, or k, should be summed over. Context should provide some clue. For example, if the sum contains energy eigenvalues, of which there generally are an infinite number, it is probably safe to assume that the summation contains an infinite number of terms. If the index is over harmonic oscillator states, which are usually numbered beginning with 0, start with 0; for the classic particle in the box problem, where a quantum number of 0 is not allowed, start with 1.

This is not to say that sums in physical chemistry never start with numbers other than 1 or 0, but if no starting value is given, you won't often go wrong by picking either 1 or 0. If you feel tempted to cast aspersions on the ancestors of your compactor, just be glad he or she is not a fan of Einstein summation notation, where the summation symbol is left off altogether and the reader is to assume that any index that appears more than once should be summed over with limits to be determined by context!

Useful Relations

Here are some things that can greatly simplify your life with sums (note that I've used various compact forms of the notation!):

$$\sum_i x_i = \sum_j x_j$$

Changing the symbol you use to refer to the index doesn't change the sum! You could use a ☺ for the index and that would work just as well.

$$\sum (x_i + y_i) = \sum x_i + \sum y_i$$

The limits on the two sums on the right must be the same as the limits on the left.

$$\sum a x_i = a \sum x_i$$

Works where a is a constant that doesn't depend on the value of the index i. If the value of the constant does change with i, then you can't pull it out of the summation.

$$\sum_{i=1}^{N} a = Na$$

Watch the index on this one. If i starts at 0, then the value is $(N+1)a$.

$$\left(\sum_i x_i\right)\left(\sum_i y_i\right) = \sum_i \sum_j x_i y_j$$

$$\neq \sum_i x_i y_i$$

Note that you have to move to two indices and a double sum here. If you don't believe this one, try it with two three-term sums!

$$\left(\sum_i x_i\right)^2 = \sum_i \sum_j x_i x_j \neq \sum_i x_i^2$$

This follows from the relation above. Again you get a double sum. Assuming equality here is one of the most common errors I see students make when they work with products of sums.

> **Example 2.1** Expand the following expression into two sums, pulling out any constants.
>
> $$\sum_J (2J+1)e^{-\beta hcBJ(J+1)}$$

Nested Sums of Series

As the last two relations above show, sums can be nested. That is, you can generate a list of sums and sum them up, essentially taking the sum of a sum. Once again, the compactness of the notation may vary, so you may have to assume values for the limits on the basis of the context. The basic notation is

$$\sum_{i=start}^{finish} \sum_{j=start}^{finish} () = \sum_{i=start}^{finish}\left(\sum_{j=start}^{finish} ()\right)$$

For example:

$$\sum_{i=2}^{4}\sum_{j=1}^{3} a_i x_j = a_2 x_1 + a_2 x_2 + a_2 x_3 + a_2 x_4 + a_3 x_1 + \cdots$$

Note that the limits don't have to be the same on the summation over i and the summation over j.

Other Forms You May See

The two indices (typically called i and j) are not necessarily independent. For example,

$$\sum_i \sum_{j>i} (\ ) = \sum_i \sum_{j=i+1} (\ )$$

Useful Relations

$$\sum_i \sum_j (\ ) = \sum_j \sum_i (\ )$$

Be sure to switch the limits as well.

$$\sum_i^N \sum_j^M (x_i + y_j) = M \sum_i^N x_i + N \sum_j^M y_j$$

The trick to seeing that this one is true is to split it into two summations and then realize that each is just a single sum done N (or M) times

$$\sum_i^N \sum_{j \geq i}^N x_i x_j = \tfrac{1}{2} \sum_i^N \sum_j^N x_i x_j$$

This shift often allows one to combine terms conveniently or to make clearer what must be done!

> **Example 2.2** Evaluate
>
> $$\sum \sum \left(\int \psi_i^* \psi_j d\tau \right)$$
>
> where $\int \psi_i^* \psi_j d\tau$ is 1 when $i = j$ and 0 when $i \neq j$.

Simple Products of Series

Here the terms of the series are multiplied rather than added. The basic notation is structured similarly:

$$\prod_{i=start}^{finish} (\ )$$

> **Example 2.3** Use the product notation to write a representation of the factorial (!) notation.

Convergence

The compactness of the summation and product notation can easily disguise the major difficulty of dealing with them: They may contain an infinite number of terms. Generally, for summations, this is handled in two ways. You can show rigorously, using a mathematical proof, that the series converges — that is, that the sum of some finite subset of terms is nearly equal to the full sum. Alternatively, you can just hope that it does and cut off the sum at whatever point seems practical to you. Usually the smallest number of terms you can get away with is three, but in some cases two will suffice.

> **Example 2.4** Would you expect the expression
>
> $$\sum_i e^{-\beta \varepsilon_i}$$
>
> to converge? Take β to be a constant and ε_i to be positive and $\varepsilon_i+1 > \varepsilon_i$. In other words, the magnitude of ε_i increases with i. Take ε_i to be $ih\nu$ (h is Planck's constant and ν is 7.5×10^{12} Hz; β is $2.430 \times 10^{20}/J$). When would you truncate the summation?

Commonly Encountered Series

A number of series recur in physics, chemistry, and mathematics. Lecturers frequently pull them out of a hat, particularly in quantum mechanics.

Common Summations

The values of certain sums can be computed just from the values of the limits. Some of these appear in statistical mechanics in particular. Of course, none of them appear in exactly these forms; rather, they are disguised as something much more complex!

$$\sum_{i=1}^{N} i = \frac{N(N+1)}{2} \qquad \sum_{i=1}^{N} i^2 = \frac{N(N+1)(2n+1)}{6} \qquad \sum_{i=0}^{\infty} x^{-i} = \frac{1}{1-x}$$

Expansions

These are series that can be used to replace functions. This can be an advantage when the function isn't known exactly, but some things are known that will allow you to compute at least some of the terms in the series. It can also be useful when the functional form of the series is easier to deal with than the true function. These occasions will arise most often in quantum mechanics and statistical mechanics.

Exponential and Trigonometric Functions

Both exponential and trigonometric functions can be represented by sums. These are often used when an algebraic form will be simpler to handle than the exact function. You can construct these forms yourself, using the Maclaurin series discussed below.

$$e = 1 + \frac{1}{1!} + \frac{1}{2!} + \frac{1}{3!} + \cdots = \sum_{j=0}^{\infty} \frac{1}{j!}$$

$$e^x = 1 + x + \frac{x^2}{2!} + \frac{x^3}{3!} + \cdots = \sum_{j=0}^{\infty} \frac{x^j}{j!}$$

$$\sin x = x - \frac{x^3}{3!} + \frac{x^5}{5!} - \frac{x^7}{7!} + \cdots = \sum_{j=0}^{\infty} (-1)^{2j} \frac{x^{2j+1}}{(2j+1)!}$$

$$\cos x = 1 - \frac{x^2}{2!} + \frac{x^4}{4!} - \frac{x^6}{6!} + \cdots = \sum_{j=0}^{\infty} (-1)^{2j} \frac{x^{2j}}{(2j)!}$$

Example 2.5 Find an expression for

$$\sum_{i=0}^{\infty} e^{-i\beta}$$

Power Series

You use these all the time in physical chemistry, though you may not be aware of it. When you fit something to a polynomial (in the simplest case, to a line), you are assuming that the function can be expanded in a power series in the region where you have data. The coefficients can be determined by applying some sort of fitting scheme (such as least squares) or by solving some ancillary equation (this approach is frequently used in methods to solve differential equations; see the appropriate section below). The general form is

$$f(x) = \sum a_i x^i$$

where the a_i are the coefficients. Obviously if you truncate this after $i = 1$, you get a line:

$$f(x) = a_o + a_1 x$$

You can obtain quadratic (*i* truncates at 2), cubic (*i* = 3), quartic (*i* = 4), or higher-order fits as well.

Example 2.6 The following energy and distance data for HCl were obtained using quantum mechanical techniques. The fundamental vibrational frequency can be found by using the curve obtained when the second derivative of the energy is plotted against displacement. Plot energy as a function of distance. Fit the data to a quadratic and to a cubic functional form. What are the values of the second-order coefficients in each case? What is the value of the HCl distance at the minimum in each case?

HCl Distance (Å)	Energy (kcal/mol)
1.0	79.620
1.1	60.690
1.2	7.220
1.3	0.000
1.4	3.018
1.5	12.350

Taylor Series

A Taylor series is just a special case of a power series where the coefficients have a particular form that depends on the derivatives of the function. Taylor series allow you to create a new form for a function in some small neighborhood around a base point x_0. They can be used to approximate a function when some of the derivatives can be determined or to simplify a function when the functional form is known. The general form is

$$f(x) = \frac{f(x_o)}{0!} + \frac{(x - x_o)f'(x_o)}{1!} + \frac{(x - x_o)^2 f''(x_o)}{2!} + \frac{(x - x_o)^3 f'''(x_o)}{3!} + \cdots$$

where

$$f'(x) = \frac{df}{dx}$$

A common place to truncate this infinite series is after the second (or quadratic) term.

Example 2.7 Using a Taylor series, expand e^{-x^2} in the neighborhood of $x = 2$ to second order and then to fourth order. Plot the two series you generated and compare them to the actual function.

Maclaurin Series

This is just a specific form of the Taylor series in disguise, where $x_0 = 0$. Starting with this form, you can, without a lot of effort, create infinite summations to represent common functions such as tangent or exponentials. Some of these were given above.

Fourier Series

This appears frequently in spectroscopic applications. The general form is

$$f(x) = \sum_{n=0}^{\infty} a_n \cos(nx) + \sum_{n=0}^{\infty} b_n \sin(nx)$$

Example 2.8 In this case, a picture is worth a thousand words. Show that you really can model any function, even if it isn't particularly well behaved, by constructing a Fourier series to represent a square wave of the form

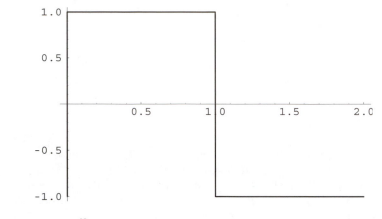

The Fourier coefficients are

$$a_n = 0$$

$$b_n = \frac{2}{L} \int_0^L \sin(\tfrac{n\pi x}{L}) dx$$

$$= \frac{4}{n\pi} \sin^2\left(\tfrac{1}{2} n\pi\right) = \begin{cases} 0 & n \text{ even} \\ \frac{4}{n\pi} & n \text{ odd} \end{cases}$$

Hints for Working With Series

- The basic hint for working with series is to be sure you know what you are working with. If you have a complex set of summations or products, a good rule of thumb is to write out the first few terms explicitly to be sure you are doing what you really want. The second hint is that if you want more than a couple of terms, it's usually worthwhile to set up a spreadsheet or a symbolic algebra package (such as Mathematica) to do it for you. The makes adding another term nearly painless, and it may help you to catch sums that are slow to converge. When in doubt — expand it out!

- Be careful when applying functions to series. For example,

$$e^{\sum x_i} = \prod e^{x_i}$$

not

$$\sum e^{x_i}$$

whereas

$$\log\left(\prod x_i\right) = \sum \log x_i$$

- On the other hand, sometimes a cigar is just a cigar[7]:

$$\frac{d}{dy}\left(\sum x_i\right) = \sum \frac{d}{dy}(x_i)$$

2–4 Logarithms and Exponentials

Way back in ninth or tenth grade, your math teacher taught you about logarithms, known to one and all as "logs", and exponentials. Not having any use for these things in this day and age (unless you are a Luddite, a calculator phobe or a slide rule aficionado), you did what any reasonable human being would do when the course was over — you forgot about them! Unfortunately, they keep showing up to taunt you, and here they are again. So for those who need a quick review of what you can and what you can't do with logs and exponents, here's a trip down memory lane.

Logs

The relationship $\log_n x = a$ is defined such that n^a is x. It is read as "log base n of x equals a". The so-called antilog of a is simply found by raising the base (n) to the power of a. The bases commonly encountered in chemistry and physics are base 10 and base e, where e is Euler's constant (2.71828...). Log base 10 is formally denoted by $\log_{10}$ but typically is written simply as log. Log base e, also called the natural log, is again formally written $\log_e$ but is more often written as ln. Just to confuse the issue, however, some folks will use log to mean natural log. Mathematica, for example, uses Log[ ] to mean natural log. To get log (base 10), you must specify the base. A handy rule of thumb is that, in quantum mechanics, log will usually mean natural log; the same is true in statistical mechanics. If the topic is chemical kinetics, it could be either.

Here are some general rules about the handling of logs:

$$\log(ab) = \log a + \log b$$

$$\log\frac{a}{b} = \log a - \log b$$

$$\log a^b = b \log a$$

Note that

$$\log(a + b) \neq \log a + \log b$$

This is a common error. Be creative — find some other mistake to make!

---

[7] If you don't get the joke here, talk to a friend majoring in psychology!

Exponentials

An exponential function is just the antilog of a natural log — in other words e^x or just 2.71828^x. You can manipulate it just as you would any other number raised to a power. See the subsection "Powers and Roots" in Section 2–12 if you need to be reminded of the details.

2–5 Trigonometry

Here are a few more things unearthed from your high school days. Remember that angles can be measured in terms of *radians* or *degrees*; π radians is $180°$. Be particularly careful when using your calculator. If the calculator is expecting angles in radians and you input degrees, your answer will (obviously) be incorrect.

Definition of the Trigonometric Functions

Classically, the trig functions are defined in terms of a right triangle. Given

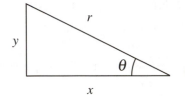

then

$$\cos\theta = \frac{x}{r} \qquad \sin\theta = \frac{y}{r}$$

The remaining functions are defined in terms of the sine (usually written sin) and cosine (usually written cos).

$$\tan\theta = \frac{\sin\theta}{\cos\theta} \qquad \cot\theta = \frac{\cos\theta}{\sin\theta}$$

$$\sec\theta = \frac{1}{\cos\theta} \qquad \cot\theta = \frac{1}{\sin\theta}$$

The periodicity of the trig functions, as well as what to do when the value of θ is negative, can be seen more clearly when the unit circle formulation is used. The accompanying circle has a radius of 1 unit. Positive angles are measured counterclockwise from the positive x-axis. The value of the cosine at any angle is given by the x-coordinate, and the sine is given by the y-coordinate. It's clear that the values of sine and cosine repeat after $360°$.

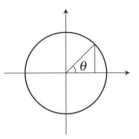

Trigonometric Identities

A few helpful trigonometric identities are

$$\cos^2\theta + \sin^2\theta = 1$$
$$\sin 2\theta = 2\sin\theta\cos\theta$$
$$\cos 2\theta = \cos^2\theta - \sin^2\theta$$

More detailed lists can be found in many calculus textbooks and in the *CRC Handbook of Mathematics.*

2–6 Derivatives

Chemistry is the study of changes in matter, so it should come as no surprise that the mathematics of change — differential calculus — plays a role. But do you remember the difference between a total differential and an exact differential? If not, read on!

The Rules

Derivatives are typically designated by $\frac{d}{dx}$, which is read as "the derivative with respect to x," or by $\frac{df}{dx}$ or by $f'(x)$. The formal definition of the derivative of a function is

$$f'(x) \equiv \lim_{h \to 0} \frac{f(x+h) - f(x)}{h}$$

More useful, probably, is a list of the derivatives of common functions:

$$\frac{d}{dx}\left(x^n\right) = nx^{n-1}$$
$$\frac{d}{dx}\left(e^x\right) = e^x$$
$$\frac{d}{dx}\left(\ln x\right) = x^{-1}$$
$$\frac{d}{dx}\left(\log_a x\right) = x^{-1}\log_a e$$
$$\frac{d}{dx}\left(\cos x\right) = -\sin x$$

Functions can be differentiated repeatedly. For example, the derivative of a function is often called its *first derivative*, and the derivative of the first derivative is called the *second derivative*.

$$\frac{d^2 f}{dx^2} = \frac{d}{dx}\left(\frac{df}{dx}\right)$$

Higher-order derivatives may be taken, though you will not see them often in chemistry.

If you have a function of more than one variable, say $f(x, y, z)$, it can be differentiated with respect to each variable. These are called *partial derivatives*. The partial derivative of a function $f(x, y)$, with respect to x at constant y, is written as

$$\left(\frac{\partial f}{\partial x}\right)_y$$

It can be computed by treating all variables except for x as constants and differentiating as usual. *Mixed partial derivatives* are higher-order derivatives of f with respect to two or more of the variables, such as

$$\frac{\partial^2 f}{\partial x \partial y}$$

As long as the function is nicely behaved, the order of differentiation doesn't matter, so

$$\frac{\partial^2 f}{\partial x \partial y} = \frac{\partial^2 f}{\partial y \partial x}$$

An *exact differential* (which is the same thing as a *total differential*) df satisfies

$$df(x, y, \ldots) = \frac{\partial f}{\partial x} dx + \frac{\partial f}{\partial y} dy + \ldots$$

This can also be written in a slightly different way as

$$df(x, y, \ldots) = A(x, y, \ldots)dx + B(x, y, \ldots)dy \ldots$$

where

$$A(x, y, \ldots) = \frac{\partial f}{\partial x} \text{ and } B(x, y, \ldots) = \frac{\partial f}{\partial y}, \text{ etc.}$$

You may be wondering why you care about this — the short answer is thermodynamics — but chances are that if you are reading this, you know why you need to know!

The Tricks

Of course, life is never as simple as one might hope, and the functions one typically must differentiate in physical chemistry are rarely as a simple as those above. Various rules can be applied to construct the derivatives of more-complex functions, including functions of more than one variable. Although the relationships that follow are written in terms of derivatives of one-dimensional functions, they are valid for partial derivatives as well.

Multiplication by a Constant

$$\tfrac{d}{dx}cf(x) = c\tfrac{d}{dx}f(x) \quad \text{where } c \text{ is a constant}$$

Sums of Functions

The derivative of the sum is the sum of the derivatives; that is

$$\tfrac{d}{dx}(f+g) = \tfrac{df}{dx} + \tfrac{dg}{dx}$$

Products of Functions (the Product Rule)

$$\tfrac{d}{dx}(f \cdot g) = f\tfrac{dg}{dx} + g\tfrac{df}{dx}$$

From this definition you can show

$$(fgh...)' = f'gh... + fg'h... + fgh'... + ...$$

Ratios of Functions (the Quotient Rule)

$$\tfrac{d}{dx}\left(\frac{f}{g}\right) = \frac{gf' - fg'}{g \cdot g}$$

Dividing Derivatives

$$\frac{\frac{dy}{dz}}{\frac{dx}{dz}} = \frac{dy}{dx}$$

$$\frac{1}{\frac{dx}{dy}} = \frac{dy}{dx}$$

The Chain Rule

Suppose you want to differentiate something that is not a product of the simple functions — for example, $\ln x^2$. The chain rule can help you accomplish this. To differentiate with respect to x, break the function down as follows:

$$x \longrightarrow (\ )^2 \longrightarrow \ln(\ )$$

The parentheses mean the contents of the previous step. Now differentiate each piece,

$$1 \longrightarrow 2(\ ) \longrightarrow \frac{1}{(\ )}$$

substitute in the parenthetical values,

$$1 \longrightarrow 2(x) \longrightarrow \frac{1}{(x^2)}$$

and multiply,

$$\frac{\partial}{\partial x}(\ln x^2) = 1 \cdot 2(x) \cdot \frac{1}{(x^2)} = \frac{2}{x}$$

"Change" of Variable

It is possible to permute partial derivatives in the following way:

$$\left(\frac{\partial f}{\partial x}\right)_y = \left(\frac{\partial f}{\partial y}\right)_x \left(\frac{\partial y}{\partial x}\right)_f$$

Another useful relation of this ilk is

$$\left(\frac{\partial f}{\partial x}\right)_z = \left(\frac{\partial f}{\partial x}\right)_y + \left(\frac{\partial f}{\partial y}\right)_x \left(\frac{\partial y}{\partial x}\right)_z$$

Example 2.9 Evaluate the following derivatives:

a. $\dfrac{-\hbar^2}{2m}\left(\dfrac{\partial^2}{\partial x^2} + \dfrac{\partial^2}{\partial y^2}\right)\left[\dfrac{2}{L}\sin\left(\dfrac{n_1 \pi x}{L}\right)\sin\left(\dfrac{n_2 \pi y}{L}\right)\right]$

b. $\dfrac{\partial}{\partial \beta}\left(\dfrac{1}{1 - e^{-\beta\varepsilon}}\right)$

c. $\dfrac{\partial}{\partial \beta}\left(\ln\dfrac{1}{1 - e^{-\beta\varepsilon}}\right)$

The Uses

One of the more obvious uses of derivatives in chemistry is to describe the rates of chemical reactions, where the rate of production of a species is given by $\frac{d[\text{species}]}{dt}$. Another spot where derivatives commonly crop up is in quantum mechanics; the

Schrödinger equation generally includes one or more derivatives in its formulation. For further information on these uses, see Section 2–8 on differential equations.

Determining Minima and Maxima

The first and second derivatives can be used to locate and characterize critical points of functions. Critical points can be maxima (mountain peaks in three dimensions), minima (valleys in 3D), or saddle points (mountain passes). All critical points are characterized by having a first derivative equal to zero. Critical points that have all positive second derivatives are minima; having all negative second derivatives indicates that the point is a maximum. Saddle points have some negative second derivatives, and some positive. In three dimensions, a saddle point will have one negative second derivative and one positive second derivative.

> **Example 2.10** Find the minimum in the radial 2s wavefunction
>
> $$R_{2s}(r) = \frac{1}{\sqrt{8}}\left(\frac{Z}{a_o}\right)^{3/2}\left(2 - \frac{Zr}{a_o}\right)e^{-Zr/2a_o}$$

2–7 Integral Calculus

This is a critical tool for physical chemistry. If you can't integrate, you can't pass! Be sure you've taken all the calculus recommended before you start, and when you can't remember what a path integral is at 2 a.m., this section should get you back on track.

The Rules

Basically, the integral represents an area. The type of integral that appears in physical chemistry is a *Riemann* integral (for a more-technical definition of this, see a good calculus textbook or consult your mathematics professor). Riemann integrals come in two flavors — *definite* and *indefinite*. Definite integrals have *limits of integration* — for example,

$$\int_a^b f(x)dx$$

Indefinite integrals do not. The definite integral can be computed from the indefinite integral as follows. Take

$$\int f(x)dx = F(x)$$

Then

$$\int_a^b f(x)dx = F(x)\big|_a^b = F(b) - F(a)$$

where the vertical bar notation is a way to say compactly "take the difference between $F(x)$ evaluated at b and $F(x)$ evaluated at a".

Integrals of Some Common Functions

$$\int x^a dx = \frac{x^{a+1}}{a+1} + C, \text{ where } a \neq -1$$

$$\int \frac{1}{x} dx = \log x + C$$

$$\int e^x dx = e^x + C$$

$$\int \sin x dx = -\cos x + C$$

$$\int \cos x dx = \sin x + C$$

$$\int \tan x dx = -\log \cos x + C$$

$$\int \log x dx = x \log x - x + C$$

The C in each of these is the *constant of integration*. Generally you can ignore this, because it will cancel out when you evaluate a definite integral. It can be determined using the boundary conditions for a problem — for example, when solving differential equations (see below).

Multidimensional integrals are evaluated in much the same way as the integrals of one-dimensional functions evaluated above. Multiple integral signs are used. For example, a three-dimensional integral would be

$$\iiint F(x,y,z)dxdydz$$

These can be definite or indefinite. Definite multiple integrals may have different limits of integration for each variable, so be alert when evaluating them! If the function being integrated is *separable* — that is, if $F(x, y, z)$ can be written as a simple product of functions of each of the variables — then the multiple integral is easy to evaluate:

$$\iiint F(x,y,z)dxdydz = \iiint f(x)g(y)h(z)dxdydz = \int f(x)dx \times \int g(y)dy \times \int h(z)dz$$

If the function isn't separable, then the integration needs to be done stepwise:

$$\iiint F(x,y,z)dxdydz = \int \left[\int \left[\int F(x,y,z)dx \right] dy \right] dz$$

where the integration over x is done first, holding y and z constant, the resulting expression is integrated over y, holding x and z constant, and finally that expression is integrated over z, holding x and y constant. (If this sounds unfamiliar and confusing, see the example below!) Remember that

$$\iiint F(y,z)dxdydz \neq \iint F(y,z)dydz$$

$$\iiint F(y,z)dxdydz = \int dx \times \iint F(y,z)dydz = x\iint F(y,z)dydz$$

Example 2.11 Evaluate the following integrals.

a. $\displaystyle \langle r \rangle_{1s} = \frac{1}{\pi a_o^{\,3}} \int_0^\infty \int_0^\pi \int_0^{2\pi} r^3 e^{-2r/a_o} \sin\theta \, d\phi \, d\theta \, dr$

b. $\displaystyle \int_0^4 \int_0^3 e^{xy} \, dx \, dy$

The Tricks

Multiplication by a Constant

$$\int a f(x)dx = a \int f(x)dx$$

Sums of Functions

$$\int (f(x) + g(x))dx = \int f(x)dx + \int g(x)dx$$

Products and Ratios of Functions

Unlike the case with derivatives, there are no fast and easy rules about what to do with an integral of a product or a ratio. Application of some of the techniques below, or recourse to numerical integration, will generally be necessary.

Integration by Parts

This lets you break up an integral of a product into (what you hope will be) simpler integrals. The key is the choice of the division. Sometimes more than one division is possible, only one of which works well. Be persistent! The general identity is

$$\int u\,dv = uv - \int v\,du$$

Typically you should choose v so that it is easy to integrate and u so that it is easy to differentiate. This is easiest to see in an example.

Example 2.12 Evaluate the following integral.

$$\int x e^{-2x} dx$$

Using a Table of Integrals

All of these tricks are often unnecessary. A good *table of integrals* is essential for survival in a physical chemistry class. Although a brief list of already-integrated expressions (both definite and indefinite) is included as an appendix in many physical chemistry texts, not all the integrals you need are covered there. A better list appears in the *CRC Handbook of Mathematics,* which your library should have. My feeling is that you shouldn't waste time using the techniques above to integrate functions if the integral has already been done. Spend your time on the physical chemistry! (Your professor may have different ideas, so check to be sure that using an integral table is OK.) To use these tables, you need to be adept at *changing variables in the integral.* For example, if you needed to know

$$\int_{x_1}^{x_2} x \log ax\, dx$$

you could browse the table of integrals and find a similar form:

$$\int x \log x = \frac{x^2}{2} \log x - \frac{x^4}{4} + C$$

The trick here to change the variable of integration from x to u, where

$$u = ax$$

$$\frac{u}{a} = x$$

$$du = adx$$

$$\frac{1}{a}du = dx$$

Substituting in the expressions above for dx and x into our target integral yields

$$\int_{u_1 = ax_1}^{u_2 = ax_2} \left(\frac{u}{a}\right) \log u \left(\frac{1}{a}du\right)$$

Note that the limits of integration have been changed! Pulling out the constants yields

$$\frac{1}{a^2} \int_{u_1 = ax_1}^{u_2 = ax_2} u \log u\, du$$

Now we can just substitute in for the integral, using the expression given in the table to get

$$\frac{1}{a^2}\left(\frac{a^2 x_2^2}{2} \log ax_2 - \frac{a^4 x_2^4}{4} - \frac{a^2 x_1^2}{2} \log ax_1 + \frac{a^4 x_1^4}{4}\right)$$

What's a Gamma (Γ) Function?

This question often arises when one is perusing a table of integrals. For example, the value of the definite integral

$$\int_0^1 x^{m-1}(1-x)^{n-1}dx$$

is given as

$$\frac{\Gamma(m)\Gamma(n)}{\Gamma(m+n)}$$

Now what? The gamma function is essentially the extension of the concept of a factorial from integers to real and complex numbers. If m and n above were integers, then the evaluation of the integral would be quite simple, because

$$\Gamma(j) = (j-1)!$$

when j is an integer. The gamma function for a real number $x > 0$ can be found using

$$\Gamma(x) = (x-1)\int_0^\infty u^{x-2}e^{-u}du$$

For example, $\Gamma(1.50) = 0.88623$. In the old days, tables of the gamma function could be found. Nowadays, it's built into many symbolic algebra packages. In Mathematica, for example, it is **Gamma[x]**.

Numerical Integration

As long as you have a function that you believe to be continuous over the interval of interest, and you have values for the function either as discrete data points or from an analytical function, you can get a numerical estimate of the value of the integral over the interval. In an older textbook, I once encountered a practical, relatively low-tech method for numerical integration of data. The authors suggested plotting the data carefully on a piece of high-quality graph paper and then cutting out the area under the curve and weighing it on an analytical balance. By subsequently weighing a piece of paper of known area (a square), one could then determine the area under the curve. These days, most students have access to much more sophisticated methods. Two simple methods, easily implemented using either paper and pencil or a computer program, are the *trapezoidal rule* and *Simpson's rule*. To use the trapezoidal rule, plot the function over the desired interval (*a* to *b*). Divide the abscissa into *n* equal intervals of length *h,* and evaluate the function at each of these points. The value of the integral can then be estimated using

$$\int_a^b f(x)dx \approx \frac{h}{2}[f(a) + 2f(a+h) + 2f(a+2h) + \ldots + f(b)]$$

This is equivalent to summing the areas of the inscribed trapezoids shown in the accompanying figure.

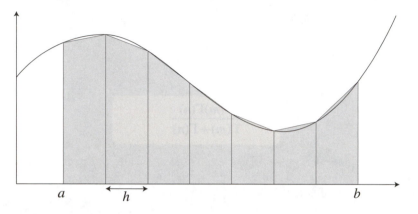

Obviously, the smaller the intervals chosen, the closer the estimate will be to the true value.

Simpson's rule works similarly. Again, divide the abscissa over the desired range into equal intervals of length h. Then

$$\int_a^b f(x)dx \approx \frac{h}{3}[f(a) + 4f(a+h) + 2f(a+2h) + 4f(a+3h) + \ldots + f(b)]$$

Many of the symbolic algebra packages, such as Mathematica and Maple, include numerical integration schemes. For example, **NIntegrate[x^2 Exp[a x], {x,0,12}]** will provide a numerical estimate for

$$\int_0^{12} x^2 e^{-ax} dx$$

See Chapter 3 for further details on using these programs.

2–8 Differential Equations

You could take an entire course on differential equations. You could, in fact, spend your life working on the solution of differential equations — I spend a significant fraction of my day doing just that! In physical chemistry you encounter differential equations in nearly every topic, including thermodynamics, chemical kinetics, chemical dynamics, spectroscopy, and, of course, quantum mechanics. Although you're not expected to be able to solve every differential equation that you can create, a passing familiarity with common techniques used to solve these equations doesn't hurt (much).

What is a differential equation? It's an equation in which both the function and its derivatives appear. Ordinary differential equations (or ODEs) contain only ordinary derivatives, whereas partial differential equations (PDEs) contain partial derivatives. If only first-order derivatives appear, then it's a first-order equation. Homogeneous differential equations can be arranged such that one side is equal to zero. An example of a first-order ODE is

$$\frac{d[O_3]}{dt} = k[O_3]$$

where $[O_3]$ is a function of time.

Systems of Differential Equations

A system of differential equations is any set of differential equations that must be simultaneously satisfied. These systems are referred to as *coupled* if the functions being solved for appear in more than one equation. You generally encounter these in chemical kinetics, where you write a differential equation for each of the species involved.

Boundary Conditions

Generally when you solve a differential equation, you produce a family of solutions. The specific member of the family that solves your problem is determined by applying known boundary conditions. Boundary conditions are generally of two types: initial conditions and multiple-point conditions. The former use the initial value of all of the variables to establish the final form of the solution. Kinetic equations generally use initial-value boundaries. Multiple-point conditions, as the name implies, specify values for the solution at several points. A good example of this type is the boundary conditions used in the "one-dimensional particle in the box" problem in quantum mechanics. Here the solution to the Schrödinger equation (the differential equation) must have the value zero at both ends of the box.

Simple Integration

If you're lucky, a differential equation can be solved simply by integrating. In the case above, rearrange the equation to get

$$\frac{d[O_3]}{[O_3]} = kdt$$

Now integrate both sides.

$$\int_{initial[O_3]}^{[O_3]_t} \frac{d[O_3]}{[O_3]} = \int_{t_{initial}}^{t} kdt$$

Note that you integrate over the boundary conditions: the initial concentration of ozone and the initial time. The integration then gives you the concentration of ozone as a function of time.

$$\ln\frac{[O_3]_t}{[O_3]_{initial}} = k(t - t_{initial})$$

$$[O_3]_t = [O_3]_{initial}\, e^{k(t-t_{initial})}$$

The key here is to be sure that you can separate the variables. In other words, can you get df and f all on one side and the independent variable, say x, on the other?

Example 2.13 A differential equation describing the concentrations of two species A and B that are isomerizing

$$A \xrightarrow{k_1} B$$

$$B \xrightarrow{k_2} A$$

is

$$\frac{d[B]}{dt} = k_1\{[A]_o - [B]\} - k_2[B]$$

Find an expression for [B] as a function of time. Assume that [B] at time $t = 0$ is zero.

Using an Auxiliary Equation

Many of the differential equations that appear in quantum chemistry can be written in the form

$$\frac{d^2 f(x)}{dx^2} + p\frac{df(x)}{dx} + qf(x) = 0$$

where p and q are constants (that is, they don't depend on x). These are examples of homogeneous second-order ordinary differential equations with constant coefficients. The solutions are easy to find once you've written the equation down in this form. The key is usually noticing that you *can* use this form. To create the solution, first form the auxiliary equation

$$s^2 + ps + q = 0$$

Solve the auxiliary equation for its two roots, s_1 and s_2. The solution to the original differential equation is then

$$f(x) = c_1 e^{s_1 x} + c_2 e^{s_2 x}$$

where c_1 and c_2 are constants that are determined by the boundary conditions.

Example 2.14 The Schrödinger equation for the particle in a one-dimensional box is

$$\frac{-\hbar^2}{2m}\frac{d^2\psi(x)}{dx^2} = E\psi(x)$$

Rearrange this appropriately and write the auxiliary equation. Write an appropriate solution.

Power Series Solutions

Another general method for solving differential equations relies on expanding the unknown function in a power series (see Section 2–3). The derivatives are then easily evaluated. This technique is best applied when there are coefficients in the equation that are not constant. An example might be the Schrödinger equation describing the harmonic oscillator, where the potential depends on the independent variable x.

$$-\frac{\hbar^2}{2m}\frac{d^2\psi(x)}{dx^2} - \alpha^2 x^2 = E\psi(x)$$

To see the power series in action, try solving the following second-order homogeneous differential equation.

$$\frac{d^2 f(x)}{dx^2} + cf(x) = 0$$

First expand the unknown function f in a power series

$$f(x) = a_o + a_1 x + a_2 x^2 + \cdots = \sum_{i=0}^{\infty} a_i x^i$$

Now find the second derivative of the power series

$$\frac{df(x)}{dx} = a_1 + 2a_2 x + 3a_3 x^2 + \cdots = \sum_{i=1}^{\infty} i a_i x^{i-1}$$

$$\frac{d^2 f(x)}{dx^2} = 2a_2 + 6a_3 x + 12a_4 x^2 + \cdots = \sum_{i=2}^{\infty} i(i-1) a_i x^{i-2}$$

Substitute the power series into the original differential equation, and voilà — no more derivatives!

$$\sum_{i=2}^{\infty} i(i-1) a_i x^{i-2} + c\sum_{i=0}^{\infty} a_i x^i = 0$$

Now all we should have to do is find all the a_i and we will know f. This would be simpler if there were only one sum, but we can't combine them unless the indices run over the same initial and final values. Therefore, we let

$$i = k + 2$$

Then you can show that

$$\sum_{k=0}^{\infty}(k+2)(k+1)a_{k+2}x^k + c\sum_{i=0}^{\infty}a_ix^i = 0$$

Try it — it really does work. the trick is to use $k = i - 2$ in the first index. Of course, because an index is just a dummy variable (see Section 2–3 again if you need to brush up on the vagaries of summation notation), we could also write

$$\sum_{k=0}^{\infty}(k+2)(k+1)a_{k+2}x^k + c\sum_{k=0}^{\infty}a_kx^k = 0$$

We can see now how to combine the sums.

$$\sum_{k=0}^{\infty}\left[(k+2)(k+1)a_{k+2} + ca_k\right]x^k = 0$$

The only way this equation could be true is if x is equal to 0 or if all the coefficients

$$\left[(k+2)(k+1)a_{k+2} + ca_k\right] = 0$$

Now we can solve for a_{k+2} in terms of a_k (a recursive relationship).

$$a_{k+2} = -\frac{c}{(k+2)(k+1)}a_k$$

Now provided that we know a_0 and a_1 (which we should be able to figure out from whatever boundary conditions we've been given), we can find all the rest of the as. If we're really lucky, we'll find that the summation we finally get for f is some recognizable sum (like the expansion of cosine). If this doesn't occur, we hope we can show that the series converges and that therefore we need only think about some (small) finite number of as.

Example 2.15 Find an appropriate form for the wavefunction describing a particle trapped in a one-dimensional box with an inclined floor. The Schrödinger equation is

$$\frac{-\hbar^2}{2m}\frac{d^2\psi(x)}{dx^2} + bx = E\psi(x)$$

where b is a constant giving the slope of the incline.

Solutions, Solutions Everywhere!

Many of the types of differential equations that appear in physical chemistry have already been solved. The key is knowing this. The problem is finding it out. Cultivate friendships with mathematicians. Bring them chocolate. They will help. If you expect to spend a *lot* of time solving differential equations, marry a mathematician. I enjoy having someone else around who was as excited as I was to find another use for the solutions to the Mathieu equation.

Numerical Methods

When I cover chemical kinetics, I show students that the approximate solutions the text generally presents for the kinetics of a multistep reaction can be not only quantitatively inaccurate but also qualitatively incorrect. For example, the accompanying graph compares the concentrations of species A, B, and C obtained using an approximate analytical solution (thin lines) to those obtained from a numerical solution (thick lines) to the differential equations. Note that early in the reaction, even the forms of the graphs aren't the same. Even at later times, where the approximation is expected to be "good," the concentration of the intermediate B is poorly predicted by the analytical solution.

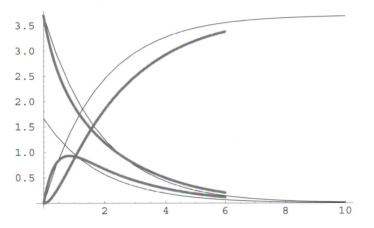

My point? You are often better off using a numerical solution! This seems odd to many students, because all of your training is to produce analytical solutions. However, given the accessibility of good numerical methods, there is no reason to live with a poor analytical solution. If the object is to train you to use physical chemistry in the real world, then you should know how to do these things. Mathematica and similar programs make it relatively simple to do this for systems of first-order differential equations.

It is useful to realize that any ODE can be reduced to a system of first-order differential equations. This can be done by introducing one or more auxiliary variables. For example, say you were trying to solve

$$\frac{d^2 f(x)}{dx^2} + x^2 \frac{df(x)}{dx} = (x-4)^2$$

You could instead create a new function z such that

$$\frac{df(x)}{dx} = z(x)$$

Then you would have a system of first-order differential equations to solve:

$$\frac{dz(x)}{dx} = (x-4)^2 - x^2 z(x)$$

which can be approached numerically.

Basically, any of these methods generates a list of the values of the function f over a range of x-values, given some initial value for the function. Instead of finding an analytical form for the function at all points and then plugging in values to generate numerical values for the function, the numerical methods jump right to the list of values. Consider a differential equation of the form

$$\frac{df(x)}{dx} = Z(x, f(x))$$

The simplest numerical method for finding the solution is Euler's method. The values of the function are predicted using

$$x_{n+1} = x_n + h$$
$$f(x_{n+1}) = f(x_n) + hZ(x_n, f(x_n))$$

where h is what is called the step size. Euler's method, although it is easy enough to do by hand, is not terribly satisfactory for most systems. You generally have to use a small step size relative to other methods, and even then, the solutions can be unstable (you are essentially assuming that the function Z is linear over the range of the step size, which generally isn't true).

The most popular method for numerically solving a system of first-order differential equations is the Runge-Kutta method. Fourth-order Runge-Kutta constructs the function as follows:

$$k_1 = hZ(x_n, f(x_n))$$
$$k_2 = hZ(x_n + \frac{h}{2}, f(x_n + \frac{k_1}{2}))$$
$$k_3 = hZ(x_n + \frac{h}{2}, f(x_n + \frac{k_2}{2}))$$
$$k_4 = hZ(x_n + h, f(x_n + k_3))$$
$$f(x_{n+1}) = f(x_n) + \frac{k_1}{6} + \frac{k_2}{3} + \frac{k_3}{3} + \frac{k_4}{6}$$

Varying step size can affect not only the quality of the solution, but also your ability to *find* a solution. It may require some playing around to find just the right step size for a particular problem. One can use these methods by hand for a few steps, but the obvious approach is to use a computer to do this. See Chapter 3 for further details.

> **Example 2.16** Solve the same differential equation as in Example 2–13, this time using Euler's method.
>
> $$\frac{d[\text{B}]}{dt} = k_1\left\{[\text{A}]_\text{o} - [\text{B}]\right\} - k_2[\text{B}]$$
>
> Find an expression for [B] as a function of time. Assume that [B] at time $t = 0$ is zero, that the values of k_1 and k_2 are 2 and 3, respectively, $A_0 = 7$, and follow the reaction for five time steps. Take the step size to be 0.1 time units. Compare your results with the analytical solution you arrived at earlier (Example 2.13). Try using a larger time step (0.5) to see what happens.

PDEs

You noticed — all the methods discussed above deal with solutions of ODEs! Generally, you won't see PDEs in introductory physical chemistry. The usual tactic is to take a PDE and separate it into a set of ODEs. For example, the Schrödinger equation describing a particle trapped in a two-dimensional box is

$$\frac{-\hbar^2}{2m}\left(\frac{\partial^2\psi(x,y)}{\partial x^2} + \frac{\partial^2\psi(x,y)}{\partial y^2}\right) = E\psi(x,y)$$

If we assume that $y(x,y)$ is separable — that is, that

$$\psi(x,y) = f(x)g(y)$$

then we can write (after a bit of algebra)

$$\frac{-\hbar^2}{2m}\left(\frac{1}{f(x)}\frac{\partial^2 f(x)}{\partial x^2} + \frac{1}{g(y)}\frac{\partial^2 g(y)}{\partial y^2}\right) = E$$

$$\frac{-\hbar^2}{2m}\frac{1}{f(x)}\frac{\partial^2 f(x)}{\partial x^2} = E - \frac{-\hbar^2}{2m}\frac{1}{g(y)}\frac{\partial^2 g(y)}{\partial y^2}$$

$$\frac{-\hbar^2}{2m}\frac{1}{f(x)}\frac{\partial^2 f(x)}{\partial x^2} = \text{something that doesn't depend on } x$$

$$\frac{-\hbar^2}{2m}\frac{1}{f(x)}\frac{\partial^2 f(x)}{\partial x^2} = \text{constant}$$

Now we have an ODE we can solve! If you need to go beyond this, it is time to take a class in differential equations.

2-9 Probability and Statistics

There are three kinds of lies: lies, damned lies, and statistics.
— Benjamin Disraeli (1804–1881)

Probabilities can be slippery things — as can their analysis (statistics). Even the "official" definition of probability is up for grabs. Fortunately, none of this should affect your use of statistics in physical chemistry. Concepts from probability and statistics show up in statistical mechanics (what a surprise!), in quantum mechanics, and to some extent in chemical kinetics. It's not hard to learn enough probability and statistics to get by. My graduate school roommate borrowed a book from me and learned enough in a couple of weeks to get out of taking probability and statistics in medical school! So here goes the really short course…

I'm going to define probability loosely as the likelihood of some outcome when there is more than one possibility. If you have n equally likely outcomes, then the probability of any one of them occurring in a given instance is

$$\frac{1}{n}$$

For example, there are 10^7 possible 7-digit phone numbers. The probability of you guessing my phone number is 1 out of 10^7. Some outcomes may be more probable than others. For example, the probability of picking up a blue M&M in a bag of peanut M&Ms is 2 in 10, while the probability of selecting a green one is only 1 in 10. The probability is often expressed as the percentage of times the outcome occurs in an infinite number of trials.

The probability of two independent events occurring is the product of their individual probabilities. The probability of any two people in the class guessing my phone number is $(10^{-7})(10^{-7})$, or vanishingly small. The probability of picking two blue M&Ms out of that package you're eating is somewhat better: $(0.2)(0.2)$, a 4% probability.

Some probabilities are correlated. For example, the probability of getting a blue M&M in a package of plain M&Ms is only 1 in 10. Suppose you had a bowl that contained both plain and peanut M&Ms (go ahead — chemistry is an experimental science!) and asked for the probability of picking out a blue one. The more peanut M&Ms you have in there, the more likely you are to get a blue one. More technically, the probability of being blue is correlated with the probability of being peanut.

It's useful to be able to figure out the number of possible permutations of objects. For example, if there are W possible arrangements, all equally probable, then the probability of any one arrangement is $1/W$. Some pertinent formulas are given below.

- The number of ways in which you can arrange N distinguishable objects

$$N!$$

- The number of ways in which you can put N objects into M boxes, putting m_1 in the first box, m_2 in the second, and so on

$$\frac{N!}{\prod_{i=1}^{M} m_i!}$$

- The number of ways to put N distinguishable objects into M boxes (any number in a box)

$$M^N$$

- The number of ways to put N indistinguishable objects into M boxes (any number in a box)

$$\frac{(M+N-1)!}{(M-1)!N!}$$

Example 2.17 Consider a large bag of M&Ms and a group of three students. If you take five brown M&Ms and distribute them to the students, how many different ways can you do this? (Assume you aren't going to be fair about it!) If you take one brown, one green, one red, one blue, and one yellow, how many different ways can you do this?

2-10 Scalars, Vectors, and Matrices — Tensors, Too

Vectors and matrices show up principally in quantum mechanics and spectroscopy. Just as summation notion makes possible the easy manipulation of sums, vectors and matrices allow the organization of information in many dimensions. The advent of symbolic mathematics packages such as Mathematica and Maple has made dealing with matrices in particular much less onerous.

Scalars

Scalars are just single numbers. For example, 2 is a scalar.

Vectors

A nontechnical definition of a *vector* is that it is an ordered list of items. A vector can also be defined as the directed line connecting two points in space. The vector connecting point A to point B is the opposite of that connecting B to A. The items in the list are called *elements* of the vector or *components* of the vector. A vector is typically designated by an arrow, $\vec{a}$. The components are generally collected between parentheses, as in $\vec{a} = (a_1, a_2, \ldots, a_n)$. Vectors can have any number of components. The following paragraphs review common operations on vectors.

Vector Addition

To add two vectors they must have the same number of components. Then

$$\vec{a} = (a_1, a_2, \ldots, a_n)$$
$$\vec{b} = (b_1, b_2, \ldots, b_n)$$
$$\vec{a} + \vec{b} = (a_1 + b_1, a_2 + b_2, \ldots, a_n + b_n)$$

Vector Multiplication by a Scalar

Just multiply each component by the scalar.

$$c\vec{a} = (ca_1, ca_2, \ldots, ca_n)$$

Dot Product or Inner Product

You can think of this as multiplication of vectors, but it's not really the same thing. The dot product takes two vectors and produces a scalar value. Given two vectors $\vec{a}$ and $\vec{b}$, each having the same number of components,

$$\vec{a} \cdot \vec{b} = \sum_{i=1}^{n} a_i b_i = a_1 b_1 + a_2 b_2 + \ldots + a_n b_n$$

The dot product is commutative.

Length, Magnitude, and Norm

All these terms mean the same thing: how long the line represented by the vector is. The term *length* is sometimes confusing because it can be used to describe both the number of components a vector has and its magnitude. The magnitude of a vector, designated $|\vec{a}|$ or sometimes $\|\vec{a}\|$, can be easily calculated by using the dot product.

$$|\vec{a}| = \sqrt{(\vec{a} \cdot \vec{a})}$$

Angle Between Two Vectors

The dot product can be used to find the angle between two vectors

$$\vec{a} \cdot \vec{b} = |\vec{a}||\vec{b}|\cos\theta$$

Unit Vector

A vector of length 1. This is sometimes designated by a "hat", $\hat{a}$. Unit vectors in the Cartesian x, y, and z directions are commonly encountered.

$$\hat{x} = (1,0,0)$$
$$\hat{y} = (0,1,0)$$
$$\hat{z} = (0,0,1)$$

Orthogonal Vectors

Vectors that are perpendicular to one another. If two vectors $\vec{a}$ and $\vec{b}$ are orthogonal, then their dot product vanishes.

$$\vec{a} \cdot \vec{b} = 0$$

Cross Product

The cross product, like the dot product, resembles multiplication of vectors. Again, the resemblance is only superficial. In this case, the result is also a vector that is perpendicular to the two original vectors.

$$\vec{a} \times \vec{b} = \hat{x}(a_2 b_3 - a_3 b_2) - \hat{y}(a_1 b_3 - a_3 b_1) + \hat{z}(a_1 b_2 - a_2 b_1)$$

A useful mnemonic for this employs the determinant

$$\vec{a} \times \vec{b} = \begin{vmatrix} \hat{x} & \hat{y} & \hat{z} \\ a_1 & a_2 & a_3 \\ b_1 & b_2 & b_3 \end{vmatrix}$$

Matrices

Just as a vector appears to be an ordered list, a matrix resembles an ordered table of values. An $m \times n$ matrix has m rows and n columns. The individual elements of a matrix **A** are typically designated a_{ij}, where i indicates the row number and j the column number of the entry into the matrix. The following paragraphs review common types of matrices.

Square Matrix

A square matrix has the same number of rows as columns.

Identity Matrix or Unit Matrix

All elements on the diagonal are 1, and off-diagonal elements are 0. An identity, or unit, matrix is always a square matrix.

$$\mathbf{I} = \mathbf{1} = \begin{pmatrix} 1 & 0 & 0 & 0 & 0 \\ 0 & 1 & 0 & 0 & 0 \\ 0 & 0 & 1 & 0 & 0 \\ \vdots & \vdots & & \ddots & \vdots \\ & & & & 1 \end{pmatrix}$$

Lower (Upper) Triangular

In a lower triangular matrix, all elements above the diagonal are zero:

$$\begin{pmatrix} a & 0 & 0 & 0 \\ b & c & 0 & 0 \\ d & e & f & 0 \\ g & h & i & j \end{pmatrix}$$

In an upper triangular matrix, all elements below the diagonal are zero.

Diagonal Matrix

All off-diagonal elements are zero.

$$\begin{pmatrix} a_1 & 0 & 0 \\ 0 & a_2 & 0 \\ 0 & 0 & a_3 \end{pmatrix}$$

Symmetric Matrix

Corresponding elements across the diagonal are equal.

$$a_{ij} = a_{ji}$$

$$\begin{pmatrix} a_{11} & a_{12} & a_{13} \\ a_{12} & a_{22} & a_{32} \\ a_{13} & a_{32} & a_{33} \end{pmatrix}$$

The following paragraphs describe common operations on matrices.

Matrix Addition

Matrix addition requires that both matrices be of the same dimension. Add corresponding elements.

$$\mathbf{C} = \mathbf{A} + \mathbf{B}$$

$$c_{ij} = a_{ij} + b_{ij}$$

Matrix Multiplication

The matrices don't need to be square, or to be of the same dimension. However, the number of columns in matrix **A** must match the number of rows in matrix **B**. To multiply an $n \times m$ matrix **A** by an $m \times q$ matrix **B**:

$$C = AB$$

$$c_{ik} = \sum_{j=1}^{m} a_{ij} b_{jk}$$

The result is an $n \times q$ matrix **C**.

Determinant

The determinant of a matrix is used in a wide variety of contexts. The determinant of a 2×2 matrix is easily given as

$$\det \mathbf{A} = |\mathbf{A}| = \begin{vmatrix} a_{11} & a_{12} \\ a_{21} & a_{22} \end{vmatrix} = a_{11}a_{22} - a_{12}a_{21}$$

The determinant of a 3×3 matrix can be found using

$$\begin{vmatrix} a_{11} & a_{12} & a_{13} \\ a_{21} & a_{22} & a_{23} \\ a_{31} & a_{32} & a_{33} \end{vmatrix} = a_{11}\begin{vmatrix} a_{22} & a_{23} \\ a_{32} & a_{33} \end{vmatrix} - a_{12}\begin{vmatrix} a_{21} & a_{23} \\ a_{31} & a_{33} \end{vmatrix} + a_{13}\begin{vmatrix} a_{21} & a_{22} \\ a_{31} & a_{32} \end{vmatrix}$$

In general,

$$\det \mathbf{A} = \sum (-1)^{i+1} a_{1i} \det \mathbf{A}_i$$

where $\mathbf{A}_i$ is the **A** matrix with the first row and the ith column removed. You keep repeating the steps, breaking down each determinant, until you have a sum of 2×2 determinants to evaluate. This sounds tedious, and it is, but take heart. If you use Mathematica or a similar symbolic mathematics package, it is a one-step procedure, such as **Det[A]**. The useful properties of determinants include the following:

- Exchange of a row or column changes the sign of the determinant.
- Adding or subtracting rows or columns leaves the determinant unchanged.
- Multiplication of a row by a constant changes the determinant by the same constant.
- If a row or column is all zeros, then the determinant is zero.
- If the matrix has two identical rows or columns, then its determinant is zero.

Inverse of a Matrix

Only square matrices have true inverses. A matrix can be inverted provided that its determinant is not zero. If you have a rectangular matrix, however, you can construct a pseudoinverse. Matrix-multiplying a matrix **A** by its inverse **A**⁻¹ (or pseudoinverse) yields the identity matrix.

$$\mathbf{A}^{-1}\mathbf{A} = \mathbf{I}$$

Finding the inverse can be tricky. The inverse of a 2 × 2 matrix is given by

$$\mathbf{A} = \begin{pmatrix} a & b \\ c & d \end{pmatrix}$$

$$\mathbf{A}^{-1} = \begin{pmatrix} \dfrac{d}{\det \mathbf{A}} & -\dfrac{b}{\det \mathbf{A}} \\ -\dfrac{c}{\det \mathbf{A}} & \dfrac{a}{\det \mathbf{A}} \end{pmatrix}$$

Linear algebra texts (see Section 2–14) describe common methods for inverting larger matrices. A good mathematical physics book is another useful source of algorithms. Finally, Mathematica or Maple can come to your rescue — **Inverse[A]** or **PseudoInverse[A]**, for example.

Transpose of a Matrix

A matrix is transposed by swapping elements across the diagonal. For example, a_{15} becomes a_{51} and vice versa.

$$\mathbf{A} = \begin{pmatrix} a & b \\ c & d \end{pmatrix}$$

$$\mathbf{A}^{t} = \begin{pmatrix} a & c \\ b & d \end{pmatrix}$$

Trace of a Matrix

Add up all the diagonal elements.

$$tr \begin{pmatrix} a & b & c \\ d & e & f \\ g & h & i \end{pmatrix} = a + e + i$$

Example 2.18 The Pauli matrices,

$$S_x = \begin{pmatrix} 0 & \frac{1}{2}\hbar \\ \frac{1}{2}\hbar & 0 \end{pmatrix}$$

$$S_y = \begin{pmatrix} 0 & -\frac{1}{2}i\hbar \\ \frac{1}{2}i\hbar & 0 \end{pmatrix}$$

$$S_z = \begin{pmatrix} \frac{1}{2}\hbar & 0 \\ 0 & -\frac{1}{2}\hbar \end{pmatrix}$$

are used to represent the spin operators. Find the operator $S_+ = S_x + iS_y$. Find $S_x S_y$. Show that the complex conjugate of the transpose of S_y is equal to S_y.

Tensors

Like vectors and matrices, tensors are ordered multidimensional objects. A tensor of rank n in m-dimensional space is an object having m^n components. A scalar is then just a zeroeth-rank tensor (in any size space). A vector is a first-rank tensor. Matrices resemble second-rank tensors, and second-rank tensors are often presented in the same format as a matrix. Tensors generally don't appear in introductory p-chem, but you will see them if you go on, particularly in spectroscopy.

2-11 Fourier Transform

The Fourier transform (FT) belongs to a class of analytical methods known as transforms. The purpose of a transform is to take a data set and apply a mathematical function to it in order to make certain aspects of the data evident. Transforms have almost certainly been used by almost every person in the fields of engineering and science. If this statement seems overly broad, consider the logarithm function, one of the simplest transforms. Take the following function:

$$\left[Br^-\right] = \frac{K_{sp}}{K_d} \frac{\left[NH_3\right]^n}{\left[Ag(NH_3)_n{}^+\right]}$$

Applying the log transform to this equation yields

$$\log\left[Br^-\right] = n\log\left[NH_3\right] + \log\frac{K_{sp}}{K_d\left[Ag(NH_3)_n{}^+\right]}$$

Thus, provided that n and K_d are constants for all values of [Br⁻] and [NH₃], the solution for n and

$$\log\frac{K_{sp}}{K_d\left[Ag(NH_3)_n{}^+\right]}$$

is trivial, because the equation is now linear in log[NH$_3$]. To determine K_d, one simply applies the inverse transform (in this case the antilog) to

$$\log \frac{K_{sp}}{K_d \left[Ag(NH_3)_n^{\,+} \right]}$$

and, provided that the appropriate values are known, solves for K_d. In this example, the application of the log transform has greatly reduced the complexity of the equation, reducing it from an nth-order equation to a first-order equation.

The purpose of the FT, like that of any transform, is to simplify the problem at hand. The FT has the form

$$S(f) = \int_{-\infty}^{+\infty} s(t)e^{-i2\pi ft}\,dt$$

where $S(f)$ is the FT of the function $s(t)$. Note the FT has frequency (f) as its domain, whereas the "original" function $s(t)$ lies in the time domain. Note also that the FT is defined over all space. It can't be done on a point-by-point basis like the log transform; you need to have all the data first and then do the transform.

Physically, the FT can be viewed as a mathematical decomposition of a complicated wave form into a sum of simple sinusoidal waves. This is (a bit) clearer if you write the FT using Euler's relationship.

$$S(f) = \int_{-\infty}^{+\infty} s(t)\cos(2\pi ft)\,dt - i \int_{-\infty}^{+\infty} s(t)\sin(2\pi ft)\,dt$$

In a sense, you project the "shadow" of the time function $s(t)$ onto all possible sine and cosine waves. Consider, for example, the wave form shown below.

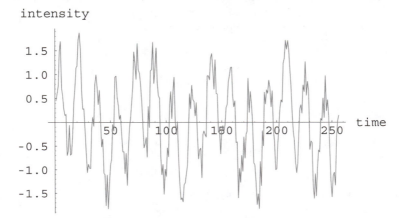

Taking the FT of this function yields the following function in the frequency domain. Note that there are two major peaks.

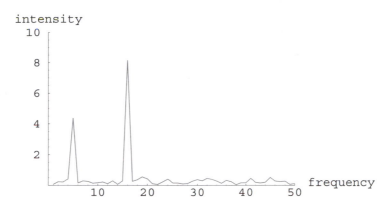

One of these peaks is at about 4 frequency units, the other at about 15. I generated the time domain plot by adding two waves together, along with a bit of "noise" for realism's sake,

$$wave(t) = \sin 30\pi t + \tfrac{1}{2}\sin 8\pi t \; + \; "noise"$$

Note that the major components of this function have frequencies of $30\pi/2\pi = 15$ and $8\pi/2\pi = 4$, so the peaks at those frequencies in the FT are not a surprise! Note also that the intensity information is preserved. The intensity of the wave with frequency 8 is half that of the high-frequency wave. The peaks in the FT show this clearly.

The so-called phase problem (something you may hear about in the context of X-ray crystallography) arises in FT for two reasons. The first is that the FT is a linear function. In other words, the FT of a combination of waves is just the sum of the FTs of the individual waves. Second, the FT is symmetric; that is, if you can transform $s(t)$ to $S(f)$, the transform of $s(-t)$ will also yield $S(f)$! This is easily seen using the above example, modified such that the low-frequency component is of opposite phase.

$$wave(t) = \sin 30\pi t - \tfrac{1}{2}\sin 8\pi t \; + \; "noise"$$

This gives rise to a slightly different wave form in the time domain.

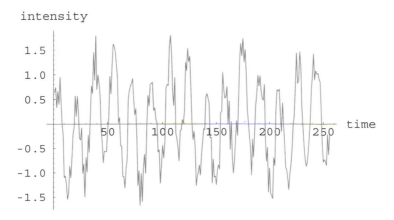

However, the FT is identical to that for the case where the low-frequency component had positive phase. In other words, you can't use the FT to tell the difference between the two waves.

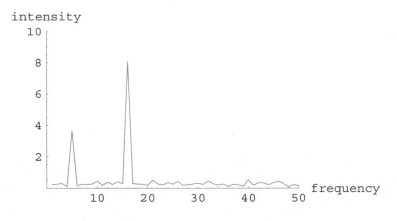

2–12 Mathematical Miscellany

All sorts of additional tidbits that might prove useful in a physical chemistry class are collected here. Some are obviously rudimentary, others a bit more esoteric.

Powers and Roots

A few basic definitions (and I mean *really* basic, so don't be insulted):

$$a^n = \prod_{i=1}^{n} a$$

$$a^{\frac{1}{2}} = \sqrt{a}$$

$$a^{\frac{1}{n}} = \sqrt[n]{a}$$

$$a^{-n} = \frac{1}{a^n}$$

$$a^{m+n} = a^m a^n$$

$$a^{m-n} = \frac{a^m}{a^n}$$

$$a^{bn} = \left(a^b\right)^n$$

Factorials

These appear in statistical mechanics and whenever you use a Taylor series. The basic definition is

$$a! = 1 \cdot 2 \cdot 3 \cdot \ \cdots \ \cdot a = \prod_{i=1}^{a} i$$

By definition 0! = 1! = 1.

Finding this number for large a can be computationally challenging; it can easily over-flow your calculator. Generally, when a is large, an estimate of the factorial can be obtained using one of the following functions:

$$a! \cong e^{-a} a^{a} \sqrt{2\pi a}$$
$$\ln a! \cong a\ln(a) - a$$

The first of these is formally Stirling's approximation and is good for $a > 10$. The second function is what is usually given in a physical chemistry text; it is good for much larger a, on the order of about 10^4. Because a is typically very large in chemical applications, on the order of 10^{14} or greater, either approximation is suitable. The second is gener-ally easier to differentiate and hence is the one used most often.

Geometry

There are all sorts of problems that require you to know such trivia as the surface area of a sphere and the volume of a cylinder. The following are a collection of commonly needed mensuration formulas. Don't put it past your professor to dream up a problem that demands something more complicated (the surface area of a cone, for example). A wider variety of mensuration formulas are collected in the *CRC Handbook of Math-ematics*. See Section 2–14 for further details.

Circles

The area A of a circle of radius r is $A = \pi r^2$. The circumference C of a circle of diam-eter d is $C = \pi d$.

Spheres

The surface area A of a sphere of radius r is $A = 4\pi r^2$. The volume V of a sphere of radius r is $V = \frac{4}{3}\pi r^3$.

Cylinders

The volume V of a cylinder of height h and radius r is $V = \pi r^2 h$.

Coordinate Systems

In general, a coordinate system is just a way to orient points or other objects in space relative to one another. The most familiar of these is probably the Cartesian coord-inate system in two dimensions. Points in this coordinate system are designated by an ordered pair of numbers *(i, j)*, where the first value (the abscissa) indicates the distance to travel along the *x* direction and the second (the ordinate) indicates the distance to travel parallel to the *y*-axis. The *x*- and *y*-axes are perpendicular to each other and intersect at the origin. And yes, you know all this. Coordinate systems, however, do not require that the axes be perpendicular. You may encounter such coordinate systems in crystallography. Some coordinate systems don't even require axes *per se* — witness polar coordinates.

Polar Coordinates

Polar coordinates are typically designated by an ordered pair *(r, θ)*, where the variable *r* gives the distance from the origin and the angle *θ* is the angle between the *x*-axis and the vector between the origin and the point described. The radial value, *r*, is always positive, and the angle *θ* ranges from 0 to 2π radians.

The relationships between two-dimensional Cartesian coordinates and polar coordinates are given by

$$x = r\,\cos\theta$$
$$y = r\,\sin\theta$$
$$r = \sqrt{x^2 + y^2}$$
$$\theta = \arctan\left(\frac{y}{x}\right)$$

The differential area element is given by $dA = r\,dr\,d\theta$.

Spherical Polar Coordinates

One can think of spherical polar coordinates as an extension of polar coordinates to three dimensions. A point is defined as an ordered triple *(r, θ, φ)*, where *r*, the radial coordinate, again specifies the distance from the origin. The azimuthal coordinate *θ* is the angle between the projection of the vector connecting the point and the origin into the *xy*-plane and the *x*-axis. The polar coordinate *φ* specifies the angle from the *z*-axis. If you didn't follow that, the figure below will probably make things a bit clearer.

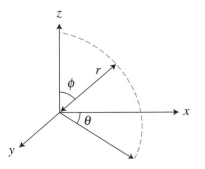

The radial coordinate is always positive. θ must be between 0 and 2π, whereas φ ranges only between 0 and π. Watch the ranges when you are integrating; a common error is to integrate both θ and φ over the range 0 to 2π. The relationships between spherical polar coordinates and Cartesian coordinates are

$$r = \sqrt{x^2 + y^2 + z^2}$$

$$\theta = \arctan\left(\frac{y}{x}\right)$$

$$\varphi = \arcsin\left(\frac{\sqrt{x^2 + y^2}}{r}\right) = \arccos\left(\frac{z}{r}\right)$$

$$x = r\cos\theta\sin\varphi$$

$$y = r\sin\theta\sin\varphi$$

$$z = r\cos\varphi$$

The differential volume element is given by

$$dV = r^2 \sin\varphi\, dr\, d\theta\, d\varphi$$

Again, a common error that students make when integrating over volume using spherical polar coordinates is to use $dr\, d\theta\, d\varphi$ as the differential volume element.

Even/Odd Functions

These functional types pop up in quantum mechanics. Typically, you first encounter them with the harmonic oscillator problem. *Even functions* are functions that are symmetric about the y-axis,

$$f(-x) = f(x)$$

Odd functions are antisymmetric about the y-axis:

$$f(-x) = -f(x)$$

Examples of even functions are x^2, e^{-x^2} and $\cos x$. Odd functions include $\sin x$ and x^3.

The product of an even function and an odd function is an odd function, the product of two even functions is an even function, and the product of two odd functions is also an even function. The most wonderful thing about these functions is what happens when you integrate them:

$$\int_{-a}^{a} even = 2\int_{0}^{a} even$$

$$\int_{-a}^{a} odd = 0$$

The result can be that all sorts of evil-looking terms are just equal to zero. Note that these relationships apply only when the limits on your integration are symmetric (–*a* to *a*)! For example,

$$\int_0^a odd \neq 0$$

2–13 Error Analysis: How Reliable Are My Results?

It's Monday afternoon and p-chem lab is in full swing across the hall. Joslyn taps on my door and brings a pile of kinetic data in with her. She pulls out a sheet and drops it on the table and asks, "Is it really zeroeth-order in iodine?" She is indirectly asking me what I suspect is the most common question in student labs: Did I do the experiment correctly? Did I get the "real" answer?

The reply seems obvious: either your value matches the one found in the chemical literature or it doesn't. If it does, you presume you did it right. If it doesn't, you can calculate how far away from the reported value you are, and that is your error. In student labs, where often one is repeating experiments that have already been reported in the literature, determining the accuracy of the result doesn't seem unreasonable and it certainly has the advantage of simplicity. But is it telling the whole story? The drawbacks of this approach are twofold. First (though you hate to think so), it may merely be luck that got you an answer that matched the literature value. There is no guarantee that you could repeat this feat. In other words, although your answer is accurate, it may not be precise. Second, this situation would rarely be encountered in a research or industrial lab. Research measurements are typically not made on systems where the values are well known. (If we knew what we were doing, we wouldn't call it research!) The critical question is really a more difficult one: How do I know my answer is right if there is no "real" value to which I can compare it? If the point of p-chem lab is to prepare you for the real-world laboratory, then we'd better prepare you to deal with this question.

Beware — what follows isn't pretty! The calculations required to assess the reliability of a numerical result can be intricate. In some cases, they can be more complicated than the calculations needed to come up with the measured value in the first place (hard to believe, I know). The widespread availability of spreadsheets and symbolic mathematical packages simplifies the process substantially. In the end, my students often find a real satisfaction in knowing that an answer that didn't appear very close to the literature value is not the result of any deficit in their technique, but is simply the best they could do within the limitations of the experimental apparatus and protocol.

Significant Figures

Just in case a year of organic chemistry has taken the edge off your ability to round numbers and use significant figures, here is a quick tour of the basics. I find I'm still writing "SF" to point out gross errors in the handling of significant figures on many exams and labs in physical chemistry; these are points that you don't have to lose (far better to lose points for things you don't know!).

- **Counting Significant Figures.** Each digit counts, except for trailing zeros used to place the decimal point implicitly. For example, 100.2 has four significant figures (SF) (the zeros in the middle are significant); 1.0×10^4 has two SF (the zero after the decimal place is significant); 0.62 has two SF (the zero before the decimal point is not significant); 1000 has either one, two, three, or four SF — it isn't clear which when the number is written this way, though most people assume the least number (1 in this case). The ambiguity can be avoided by writing the number in scientific notation.

- **Rounding.** The digit after the last significant digit is used to determine whether the number rounds up or down. If the following digit is 5 or greater, round up, otherwise round down. For example, 84.141 rounded to four SF is 84.14, while 5.246 rounded to three SF is 5.25.

- **Adding and Subtracting.** When two numbers with differing numbers of significant digits after the decimal point are added or subtracted, the final result retains only as many significant digits after the decimal point as the less accurate of the values. Example: 4.56 + 12.8197 = 17.38, not 17.3797.

- **Multiplication and Division.** The number of significant figures in a product or quotient is equal to the number in the less accurate of the combined values. For example, $0.67 \div 0.3498 = 1.9$, not 1.92 or 1.9153802.

- **Logarithms.** The sticking point here is that you may end up with more SF than you started with. The number of SF in the original number gives the total number of SF after the decimal point in the computed logarithm. For example, Log(1.381 $\times 10^{15}$) = 15.1402, not 15.14.

- **Exponents.** The reverse of logarithms (logical, no?). Use the number of SF after the decimal point. For example, $e^{2.34}$ is 10., not 10.4 or 10.381237.

- **Exact Numbers.** Some numbers are known to an arbitrary precision — for example, π is 3.14159...; the only catch is that some numbers can go either way, depending on the context. If you measure a pressure of 2 torr, then the 2 has one SF. On the other hand, if the formula you are using includes P^2, then the 2 here is exact and you may use as many SF as you need.

Above all, be sure not to write down all the digits on your calculator. Just because they are there does not mean they matter to the answer. For example, if you are offered 25% off on an item that costs $199.99, your calculator might say you should pay $149.9925; however, that last $0.0025 doesn't matter and you will pay $149.99. Of course, if you're really on top of it, you notice that 25% has only two SF, so you really should pay $150!

Errors and Mistakes Aren't the Same Thing

When I ask students to write about the error in their measurements in a lab report, the comment that appears most often is "human error", which I translate as "I might have made a mistake — and I'm trying to cover myself". In most assessments of error in the scientific literature, however, human error is never mentioned. It is presumed that if such errors were made, the experiments were thrown out and not considered in drawing the conclusions presented. So where does this leave the writer of that p-chem lab report who is required to discuss sources of error?

We can classify errors in experiments as follows:

- systematic errors,

- random errors,

- model errors, and

- blunders.

Systematic errors result when a consistent bias is present in an instrument itself, in its operation, or in a protocol or methods. For example, some manometer might consistently display a lower pressure than the actual pressure, or a reaction that you are following might have a side reaction that you are not considering. These errors are typically caught by calibrating your instruments or by carefully testing your protocol with known systems. This type of error can be factored out of your data either by recalibrating the instrument or by adjusting your data. For example, Spec21D spectrophotometers output a voltage between 0 and 1V that should be equal to the percent transmittance (%T) of the sample. However, if you measure the voltage on any given Spec21D and compare it to the displayed %T value you may find that the relationship isn't perfect. The voltage might be 0.76V when %T is 0.81. This isn't of critical importance *unless* you are using the voltage data directly — for example, if the Spec21D is interfaced to a computer. The problem can be dealt with by tweaking a little screw on the spectrophotometer until the values match — that is, by recalibrating the instrument. Alternatively, a relationship (with luck, a linear one!) can be derived between the measured value of the voltage and the displayed value of the transmittance on the basis of a series of measurements. This equation can then be used to correct your data after the fact; most data-acquisition software will even let you do it on the fly. Be sure to check the calibration of the instrument throughout the range over which you plan to use it. In the example above, the relationship of voltage to %T should be examined at the wavelength used in the measurements and over the range of %T to be measured. It is often the case that an instrument that is well calibrated for %T between 95% and 50% is poorly calibrated at 5%. If you are not going to measure in the problematic range, or if you can arrange your experiment so that you don't have to, the problem of recalibrating the instrument or adjusting your data is simplified.

Example 2.19 Given the following table of values, determine an appropriate *linear* calibration equation relating percent transmittance (T) to measured voltage (V).

Voltage	% Transmittance
0.05	0.01
0.10	0.07
0.20	0.20
0.30	0.31
0.40	0.41
0.50	0.52
0.60	0.61
0.70	0.70
0.80	0.80
0.90	0.90

Comment on the range of percent transmittances that an experimenter can expect to measure reasonably using your calibration equation. Can a reliable value for percent transmittance be determined when the measure voltage is 0.045V? When it is 0.45V?

Random errors arise from intrinsic limitations in either instrument sensitivity (generally in the case of digital instrumentation) or the intrinsic ability of the human eye to read an instrument (generally in the case of analog instrumentation). An example of the first is a digital pH meter that displays pH as ☐☐.☐. The instrumentation cannot detect (or at least it cannot report) changes in the pH of less than about 0.05 unit. An example of the second case is the reading of a mercury thermometer marked at intervals of 0.1°C. Given the limitations of the sensitivity of the human eye, it would not be possible to read this thermometer to ±0.000001°C. On the other hand, an experienced person could read it to a precision of ±0.02°C.

Example 2.20 Next time you are in the laboratory, carefully read the fine print on the glassware — say, on a graduated cylinder, a beaker, and a volumetric flask. How do the intrinsic limitations differ among these three tools for measuring volume?

Model errors aren't really errors in the sense that I've described above. These occur when your data don't meet your expectations. For example, if you were studying the pressure dependence of a gas volume at constant temperature, you would expect the volume to decrease as $1/p$. At low pressures this may be the case, but at higher pressures you are likely to find that the data does not follow your expected pattern. In this case, it could be either a true error (see the types discussed above) or a failure of your model (for example, the ideal gas law doesn't work for real gases in all situations). Lots of interesting science is discovered in this way!

Example 2.21 The following data were obtained for the isomerization of cyclo-propane to propene. The reaction is presumed to be first-order, so a plot of ln(pressure/initial pressure) versus time should be linear, and the slope of the line will yield the rate constant. Calculate the rate constant on the basis of each of the following data sets. What do you conclude about data set 1?

Data Set 1	$P_0 = 95.7$ torr, T = 498°C			
Time (seconds)	50	125	206	301
Pressure (torr)	94.7	92.9	91.2	89.3

Data Set 2	$P_0 = 196$ torr, T = 501°C			
Time (seconds)	51	122	204	297
Pressure (torr)	189	179	169	158

Data Set 3	$P_0 = 585$ torr, T = 499°C			
Time (seconds)	48	119	199	295
Pressure (torr)	565	537	506	473

Blunders are just what they sound like. These are sometimes called avoidable errors. "Avoidable" isn't meant to imply that a good scientist will never make this type of error. It *does* imply that a good scientist will avoid using the data from experiments in which these types of errors were made (or from experiments where she strongly suspects that such errors have occurred). In formal scientific writing, you rarely see this type of error discussed, because the assumption is that if you have made this sort of error (or even think you have made this sort of error), you will throw out all the data associated with the blunder and/or stop the experiment and begin again. Examples of such errors include misreading a balance (did you really use 50.0g of benzoic acid in the bomb?), dropping some of an already weighed sample on the floor, forgetting to take an initial temperature until 5 minutes after an experiment began — the list is obviously endless. When discussing potential sources of error in the formal professional report of an experiment, blunders are rarely mentioned, because it is assumed that if such an error was made, you threw out the data! (An exception might be made if the error was an uncommon one and could be dangerous.)

Estimating Errors on Single Measurements

Manufacturer Specifications for Digital Equipment

Often the manufacturer of a piece of equipment will provide (either on the instrument itself or in the accompanying manual) the precision with which the equipment makes measurements. When such information is not available, it is customary to assume that there is an error of ±0.5 in the last digit displayed by a digital display.

Analog Equipment and Glassware

Typically, one can read an analog scale to within ±0.2 or ±0.5 of the last interval marked. (For example, if the scale is marked in increments of 100, one could at best read it to ±20 units). Volumetric glassware is usually marked on the glassware itself (see Example 2.20). For very-high-precision work volumetric glassware must be calibrated, a tedious — but necessary — procedure. In this case the uncertainties in the volume measurements must be calculated statistically; see the following section.

Errors in Sets of Measurements

Presuming that most of the error in a series of experiments is due to random error (in other words, assuming that you've eliminated data arising from known or suspected blunders, calibrated your method, and corrected your data for systematic errors), statistical methods can be used to find both the final measured value and the estimated error in that value. In physical chemistry lab, this generally means either finding the mean value of a set of measurements or doing a linear fit to a data set. You can estimate the uncertainty of values determined this way; the details follow.

Calculating the Mean

To calculate the mean value <p> of a set of N values $\{p_1, p_2, \ldots, p_i, \ldots, p_n\}$ use:

$$\langle p \rangle = \frac{1}{N} \sum_i^N p_i$$

Example 2.22 The enthalpy of combustion of naphthalene was measured using bomb calorimetry. The experiment was repeated five times with the following results (all in kJ/mol): 4940, 4640, 6076, 4770, 4770. What is the mean value of the enthalpy?

Calculating the Standard Error of the Mean

To get an estimate of the precision of the mean value of a set of N values $\{p_1, p_2, \ldots, p_i, \ldots, p_n\}$ whose mean is <p>, compute the standard deviation from the mean, s, using

$$s = \sqrt{\frac{1}{N-1} \sum_i^N (p_i - \langle p \rangle)^2}$$

Then the standard error of the mean $s_{<p>}$ is given by

$$s_{\langle p \rangle} = \frac{s}{\sqrt{N-1}}$$

Example 2.23 What is the standard error of the mean for the data in Example 2.22?

Confidence Limits

The standard error of the mean is rarely used directly as a measure of the precision of the mean value. Instead, one uses a confidence limit (CL). The confidence limit gives one a statistical estimate of the range of values in which one is X% likely to find the "true value" of the quantity under consideration. Typical values of X are 90%, 95%, and 99%. You may see values reported as $A \pm a(95\%)$, which means that the true value of A is 95% likely to be found between $A - a$ and $A + a$.

To compute the confidence limit, one relies of a table of t factors (a selection is given below). To compute the error on a mean value, taking into consideration the confidence limit, use

$$error = s_{\langle p \rangle} t$$

where the value t depends on the number of values. A brief table of t factors follows (larger tables can be found in the *CRC Handbook of Mathematics* and in other statistical compilations).

Number of Values	t for 90% CL	t for 95% CL	t for 99% CL
2	6.31	12.70	63.70
3	2.92	4.30	9.92
4	2.35	3.18	5.84
5	2.13	2.78	4.60
21	1.71	2.09	2.85
31	1.70	2.04	2.75
∞	1.64	1.96	2.58

Example 2.24 Compute the error in the mean for the heat of combustion of naphthalene, assuming that a 90% confidence limit is required.

Discarding Data

Points that lie far from the mean are sometimes called *outliers*. They may be the result of an unnoticed blunder, or they may just be outliers. One method for discarding these points is simply to look at the data and decide which points "look awful". Do not try this at home. Certainly do not try it in p-chem! Instead use the following procedure:

- Identify possible outliers. In doing this, do not attempt to discard more than 20% of your data or to discard data points that are duplicates of one another.

- Recompute the mean without the possible outlier(s).

- Compute the average deviation from the *new* mean for the set of data using

$$d_{av} = \frac{1}{N} \sum_i^N |(p_i - \langle p \rangle_{new})|$$

- Test each outlier $p_{outlier}$, and if the condition below is met, discard the value.

$$p_{outlier} - \langle p \rangle_{new} \geq 4d_{av}$$

- Recompute the values of s, $s_{<p>}$, and the error on the mean, including confidence limits.

> **Example 2.25** Are any of the values given in Example 2.22 for the enthalpy of combustion of naphthalene possible outliers? Test them using the above protocol.

> **Example 2.26** What value, including error limits, should a student with the experimental results given in Example 2.22 report for the enthalpy of combustion of naphthalene?

Linear Regression

Physical chemistry frequently requires fitting data to a line or curve. Most physical chemistry courses do little to dispel the notion! Once again, one needs to be alert to the size of the error in the slope and intercept that results from the underlying errors in the raw measurements of x and y. The (x, y) regression data encountered in the physical chemistry lab generally is assumed to have an uncertainty in y that is larger than x. One also assumes that the errors in y are fixed — in other words, each y has the same error. Then the slope and intercept are given by

$$\text{intercept} = \frac{\sum_i^N x_i^2 \sum_i^N y_i - \sum_i^N x_i \sum_i^N x_i y_i}{N\sum_i^N x_i^2 - \left(\sum_i^N x_i\right)^2}$$

$$\text{slope} = \frac{N\sum_i^N x_i y_i - \sum_i^N x_i \sum_i^N y_i}{N\sum_i^N x_i^2 - \left(\sum_i^N x_i\right)^2}$$

If the errors in the y-values being plotted have the constant value σ, the errors in the regression slope and intercept can be computed using

$$\text{error in the intercept} = \sqrt{\frac{\sigma^2 \sum_i^N x_i^2}{N\sum_i^N x_i^2 - \left(\sum_i^N x_i\right)^2}}$$

$$\text{error in the slope} = \sqrt{\frac{N\sigma^2}{N\sum_{i}^{N}x_i^2 - \left(\sum_{i}^{N}x_i\right)^2}}$$

Note that these numbers don't tell you anything about how good a fit to a line you have. They report only how much the error in the dependent variable influenced the error in the slope.

> **Example 2.27** Using data sets 1 and 2 from Example 2.21, calculate the slope and the error in the slope of the best-fit lines. Assume the error in $\ln(P/P_0)$ for the first set is 0.0008 and for the second set is 0.0004. Are the errors in the rate constant comparable for the two data sets? If not, why not?

Propagation of Errors

When measurements are combined (for example, when one calculates concentration, where the mass may be known to five significant figures and the density to only two), the random errors in each will contribute to some amount of error in the final result. This is generally referred to as propagation of errors, and various methods are used to account for it. More details are available in many analytical chemistry books, but most of the basic skills are outlined below.

Addition and Subtraction

Suppose one must add or subtract two measurements, $A \pm a$ and $B \pm b$. To calculate the error c in $C = (A + B)$ or $C = (A - B)$, use

$$c = \sqrt{a^2 + b^2}$$

> **Example 2.28** Using the data in the accompanying table, calculate ΔT for each run and the error in ΔT. The error in each T is 0.05°C.
>
> | Initial T (°C) | 25.76 | 25.46 | 25.17 | 24.54 | 24.95 |
> |---|---|---|---|---|---|
> | Final T (°C) | 28.73 | 29.71 | 28.30 | 26.73 | 26.52 |

Multiplication and Division

Suppose one wishes to multiply or divide two measurements, $A \pm a$ and $B \pm b$. To calculate the error c in $C = (AB)$ or $C = A/B$, use

$$c = C\sqrt{\left(\frac{a}{A}\right)^2 + \left(\frac{b}{B}\right)^2}$$

Example 2.29 Using the first data set from Example 2.21, calculate the error in the ratios of the pressure to the initial pressure. Assume an error in the pressures of 0.05 torr. Are the errors the same for each time?

More-Complicated Expressions

Suppose one has an expression Q that is a function of $A_1 \pm a_1$, $A_2 \pm a_2$, etc. To calculate the error q in Q, use

$$q = \sqrt{\sum_i \left(\frac{\partial Q}{\partial A_i}\right)^2 a_i^2}$$

Example 2.30 Suppose you are trying to calculate the change in free energy from the equilibrium constant K for the following reaction:

$$\text{keto} \leftrightarrow \text{enol}$$

Calculate ΔG for the reaction and the error limits on it, given the data below. Recall that

$$\Delta G = -RT \ln \frac{[\text{enol}]}{[\text{keto}]}$$

The concentration of the ketone is $1.78 \times 10^{-2} \pm 2 \times 10^{-4}$M, the concentration of the enol is $7.53 \times 10^{-5} \pm 2 \times 10^{-7}$M, the temperature is 298 K $\pm$ 1 K.

2–14 More Information

Guerilla math is meant only to get you through most of the rough spots. At times you may need to review material more thoroughly or get deeper into the subject. A list of useful references follows, along with my assessment of how accessible the information in them is.

- *Applied Mathematics for Physical Chemistry* by James R. Barrante (Upper Saddle River, New Jersey: Prentice Hall, 1998). As the title suggests, this book focuses on the math you need for p-chem. There is somewhat more detail on the mathematics than I give here. The example problems generally aren't targeted toward physical chemistry, but are the sort you'd encounter in a math course.

- *Calculus* edited by Deborah Hughes-Hallet (New York: John Wiley & Sons, 1997). This is an excellent introductory calculus textbook. If you don't have a text on hand, it would be useful to find this book in the library. It approaches topics from a variety of angles, which should help the casual reader.

- *Basic Multivariable Calculus* by Jerrold E. Marsden, Anthony J. Tromba, and Alan Weinstein (New York: W.H. Freeman & Company, 1994). As the name suggests, this is a good basic, multivariable calculus textbook. Your best resource is probably the text you used when you took the course, but if you need additional help or don't have a copy, this book should ably fill the gap.

- *CRC Concise Encyclopedia of Mathematics* by Eric W. Weisstein (Boca Raton: CRC Press LLC, 1998). This can replace almost any math book in your collection. It's pretty readable and covers everything you could possibly imagine and more. A CD-ROM version is available.

- *CRC Standard Mathematical Tables and Formulae,* 30th edition, edited by Daniel Zwillinger (Boca Raton: CRC Press LLC, 1995). This is an excellent reference tool, probably worth buying if you plan to go on in physics or chemistry. It includes tables of integrals and much useful mathematical miscellany. I've been using my father's high school copy (the 9th edition!) for years. This edition is twice as long and omits the once-useful log tables.

- *Numerical Recipes: The Art of Scientific Computing* by William H. Press (various editions). This book presents various numerical techniques useful in the sciences. It does not contain a lot of in-depth information on any given method, but it is extremely readable. The strength of this book is its presentation of algorithms. Versions are available in many programming languages, including C, FORTRAN, Pascal, and BASIC, and for various platforms, including the Macintosh and parallel machines.

- *Data Reduction and Error Analysis for the Physical Sciences* by Philip R. Bevington and D. Keith Robinson (New York: McGraw-Hill Companies, 1992). This classic book provides in-depth coverage of the topics addressed in Section 2–13 as well as many other areas left untouched here.

- "The NIST Reference on Constants, Units and Uncertainty", available on the World Wide Web at http://physics.nist.gov/cuu/uncertainty/index.html.

- *Guidelines for Evaluating and Expressing the Uncertainty of NIST Measurement Results,* NIST Technical Note 1297, by B.N. Taylor and C.E. Kuyatt, 1994.

Chapter 3

Beyond Pencil and Paper

One of my neighbors, himself a chemist, used to greet me with "There's the graphite and cellulose chemist!" — he assumed that as a theoretical chemist I was spending all my time with pencil and paper. I used to joke back that I was really a "silicon" chemist, because most of my work was done on the computer! Computers are a ubiquitous part of not only theoretical chemistry, but of the whole of physical chemistry.

Computer programming is a useful skill for a student of physical chemistry, but the use of computers goes well beyond that in the typical course. Word processing is practical, since multiple drafts of reports may be required. Symbolic math programs such as Maple and Mathematica enable complex equations to be evaluated and their results displayed. Spreadsheet programs make manipulating large data sets less onerous. Data acquisition by computer in the laboratory is common. Most students in physical chemistry courses have at least some familiarity with a computer and its various uses. This chapter tries to provide enough information to get a complete novice started, as well as some basic instruction in programming for those who've never dabbled in that arcane art.

There is no way that I could cover all of the available and commonly used software, and versions change rapidly, so I've chosen to illustrate the concepts using a small subset of software. Generally, if one package can do it, so can its competitor — the trick is generally knowing that the task can be accomplished at all. Check the documentation of your particular version of software to find out how to do the task described.

No matter what type of software you use, be aware that computers are picky about many things, including (but not limited to) spaces, commas, and case. If the command required is **Log**, often **LOG** won't work. And a missing semicolon can throw off a whole section of code, creating a cascade of errors.

Even if you are very comfortable using computers, I urge you to read the section on computers and productivity that follows! I know you are taking all my motherly advice, so humor me and read it.

3-1 Using a Computer Productively

Avoid the Tom Sawyer Syndrome

Tom Sawyer, charged with painting his aunt's fence, convinces his friends that the endeavor is so much fun that they not only beg him to let them do it but also pay for the privilege. It's very easy to fall prey to the same fascination with a computer. The available software lets you do marvelous things. Like Tom's friends you may be convinced that they are fun, and you certainly are paying the software developers to make it possible to do them. Thus, it is very tempting to make a computer do incredible things just because it can. Resist this temptation. First, before getting incredibly carried away with formatting text or graphics, or with improving the input or output of your program, consider whether it's worth the time. If the text is clear, the graphic readable, and the input request understandable, don't mess with it! Second, there are generally several ways to accomplish the same task; choose the method that will cause you the least fuss. If you need to plot a data set, you could write a program, use a spreadsheet, or employ a symbolic math package. You might choose the first option if you have a previously written program you can modify. Alternatively, the use of a spreadsheet might be more attractive, particularly if you are already comfortable using it. A student once tried to convince other students (and me) that it was really much better to write your own word processor than to use a commercial product. Having written my own word-processing program to draft my Ph.D. thesis (no Macs or PCs at the time!), I certainly didn't lack the technical skill to do the job, but I couldn't figure out why anyone would spend the time when they didn't have to. Don't reinvent the wheel — let Microsoft do it for you!

Data Storage

There are two kinds of computer users: those who have lost data and those who will. The first rule of data storage is to keep a backup. If the material is really important to you, keep two backups! Computer technology changes rapidly, and some of the suggestions that follow will be obsolete or unavailable in a few years, but the underlying principle remains the same: Always back up!

Two guiding principles of data storage are accessibility and durability. For most student laboratory work, the first is the more critical. You want to be sure that in case of catastrophe, you can easily and rapidly reconstruct your data and any files you may have used to work it up. For short data sets and uncomplicated mathematical manipulations, a printout of the data that you could retype is fine. For larger data sets and more-complicated spreadsheets or symbolic math notebooks, other media (such as a floppy disk or a Zip disk) are probably more convenient. If you are sharing data in a group, be sure every person has his or her own copy of the data. Your copy can serve as a backup for others in the group, and you will not have to find someone to retrieve the data when you want to write it up (late p-chem lab excuse #216: "My lab partner went home because of a family emergency and took all my data with her").

Most people don't feel the need to keep their p-chem lab data for posterity, but permanence and durability will likely be more of an issue when you do research. Be alert to backup and disk-cleaning schedules on shared machines. There is nothing like that sinking feeling when you come back Monday morning to find out that user files are automatically wiped every Saturday night and your 10-hour ^{13}C NMR run of Friday is just so many random bits. If the machine is backed up, be sure you find out who backs it up, what the schedule is, how you get access to data, and how long will it take to get data back. (If it isn't backed up, remember the first rule of data storage.) If you are storing data for the long haul, as in your Ph.D. thesis, the durability of your storage medium is of prime importance. A magnetic tape can degrade to the point that it is unreadable in less than two years if storage conditions are not ideal. Ideal storage conditions may be available in a communal storage facility, but you run the risk of other losses. Two of my students lost some of their Ph.D. work when the management of the college storage facility changed. The new staff didn't realize that outside people were using the storage and tossed all the tapes they didn't recognize. CD-ROM can be a better medium for longer-term storage, but you may have to move your data to a machine that can write CDs. Another alternative is to keep critical data and programs on several machines, preferably machines that are regularly backed up.

3–2 How to Speak Computer: A Glossary for the Virtual Novice

In a parenthetical comment in his text *Quantum Chemistry*, Ira Levine notes that "[i]f you learn enough abbreviations you can convince some people you know quantum chemistry". The same applies to the terms tossed around by the computer savvy (or those who at least appear savvy!). Read the list and you'll get the pun in the section title.

- **ASCII A**merican **S**tandard **C**ode for **I**nformation **I**nterchange — actually a 7- or 8-bit code used to represent letters, numbers, and symbols, but the term is often used as a synonym for text. An ASCII file in this context is a collection of letters, but no "special characters" — that is, nonprinting characters such as control codes that tell the computer to do something. Word-processing documents, though they appear to be just letters, usually include nonprinting codes to control the formatting. You can usually save files in ASCII or text form.

- **backup** An extra copy of something, kept in case something happens to the original version on the computer.

- **batch** All jobs that are currently being run by the computer "behind the scenes".

- **baud** A unit specifying the rate of a data transfer; bits per second.

- **binary** Base 2. Ones and zeros. Not ASCII. Executable code is usually kept in binary.

- **bit** The smallest piece of data stored in a computer. A zero or a one.

- **boot** Tell the computer to start.

- **bug** An error or a problem.

- **bus** A piece of hardware that shuttles data around.

- **byte** 8 bits make a byte.

- **cache** An extra storage spot, usually one that can be accessed quickly. Often-used commands can be stored there.

- **chip** A little wafer of silicon in your CPU that makes the computer work.

- **client** A computer that is looking to a host for information or requesting that a task be done.

- **clipboard** A place where the last item cut or copied resides.

- **code** Instructions used by the computer to execute a task.

- **compression** A mechanism for squeezing excess space out of a data file. Compressed files must be uncompressed to be used. Compression is used when compact storage is desired or when files must be transferred — for example, downloaded.

- **CPU** The **c**entral **p**rocessing **u**nit is the guts of your computer.

- **crash** To cease to work. Computers crash. Codes crash.

- **cursor** The little line that keeps track of where you are on the screen.

- **cut/paste** To remove a piece of text or figure and reposition it (presumably elsewhere).

- **default** Computerese for factory settings. These are specifications for a computer program that someone assumed would fit most situations. Usually they can be changed.

- **directory** A list of files; a collection of files.

- **download** To move data (files or codes) from one machine to another.

- **encryption** Coding data in such a way that it can be read only by the owner or other party who has the key.

- **external** Outside the computer's case — for example, an external modem.

- **FAQ** **F**requently **a**sked **q**uestions.

- **FLOP** **F**loating-**p**oint **o**perations per **s**econd. A measure of how fast a computer really runs!

- **FTP** This stands for **f**ile-**t**ransfer **p**rotocol, but it usually means a program that uses a standard file-transfer protocol to move data from one machine to another.

- **hardware** The actual machinery that makes up a computer system; generally includes a CPU, a monitor, and various peripherals, such as printers.

- **host** A computer that is in the business of providing data to other machines, or one that executes tasks for other machines.

- **icon** A little picture on the screen. Generally, if you click on it, something will happen.

- **interactive** The opposite of batch. The computer executes your commands in real time while you wait.

- **internal** Inside the computer's case — for example, an internal modem.

- **IP** Internet protocol; as in IP address.

- **ISP** Internet service provider. Someone who has a host computer and provides time to client machines, usually to access the Internet, but also to enable other data transfers (such as FTP).

- **K** A kilobyte, or 1024 bytes (the prefix kilo implies 10^3 to the metrically inclined, but in fact a kilobyte is 2^{10} bytes, just roughly a thousand). Sometimes abbreviated Kb or Kbyte.

- **LINUX** An open-source version (that is, it contains no proprietary code) of the UNIX operating system. The code is coordinated by Linus Torvalds of Sweden. The name is derived from Linus UNIX.

- **local** On your own machine, as opposed to on a host or at a remote location.

- **log in** Provide a name and (usually) a password to obtain access to a system.

- **log out** Discontinue access to a system.

- **Mbyte** A megabyte. Technically 1024 Kbytes (1,048,576 bits), but sometimes loosely used to mean a million bytes; also written just M.

- **media** Something on which data can be stored. Paper is a medium, so is a floppy disk, and so is the hard drive in your machine. The latter two are magnetic media.

- **meg** How M (Mbyte) is pronounced.

- **memory** A mechanism for storing data. This term typically refers to RAM or ROM, but it is sometimes used to refer to amount of storage available on media.

- **modem** **Mod**ulator-**dem**odulator. A device that lets you move data over a phone line. These can be external or internal.

- **mouse** A nifty little device that lets you move the cursor around the screen. Mistaken by Star Trek's Mr. Scott for a microphone.

- **network** A collection of computers and peripherals (such as printers) hooked together via cables. Routers are used to move data to appropriate destinations.

- **nybble** Half a byte, or 4 bits.

- **operating system** Also called the OS, the computer program that tells your computer how to do basic operations, such as copy a file.

- **password** A secret word used to gain access to a system or to particular files within a system. Never, ever, *ever* tell anyone your password, even the computer-center people. You can be held responsible for any actions taken on your account if you let your password out. Good passwords aren't guessable (don't use your birthdate or your girlfriend's name or single words in any language). Include uppercase and lowercase letters as well as "non-letters" such as & and ! and be sure your password is at least eight characters long. Change passwords regularly. Some systems will enforce some or all of these guidelines. These days you can have many passwords — and it can be difficult if you can't remember one. Generally you aren't supposed to write them down, but if you're like me and have more than 20, many of which change monthly, your memory may not be equal to the task. If you *must* write them down, consider the following suggestions: Don't write a password on the same sheet of paper as your user name and/or the name of the computer for which the account is used. Don't carry the sheet around with you; leave it in your room or car. Use clues to remind you of what a current password is (even if you can't select your password, this one can work). Keep the list on your PDA and password the *list* (that way you only have to remember *one* password). Finally, know how to get access if you should forget your password!

- **path** How to get to a specific file. In UNIX this term refers to the directory specification. For instance, the path of the file containing the draft of this chapter is /HD/MF/SurvivalGuide/Chapter3.

- **PC** **P**ersonal **c**omputer. This term typically means a Wintel or Windows-based machine (as opposed to a Macintosh).

- **PDA** **P**ersonal **d**igital **a**ssistant. A tiny little computer (for example, a PalmPilot) that can keep track of data such as a calendar or address book and can do small tasks, such as pick up and send e-mail or play chess. Some PDAs are showing up in the laboratory to do data acquisition.

- **peripheral** Something that is hooked up to the computer, such as a scanner, printer, or Zip drive.

- **platform** A type of computer. What platforms do I use? Macintosh and SGI.

- **port** As a noun, a place where data enters or exits the computer. Peripherals plug into ports. As a verb, to modify code to work on another platform.

- **programming language** Instructions that tell the computer to take various actions. Examples of programming languages include C, C++, FORTRAN, and BASIC.

- **RAM** Random-access memory. The memory inside the CPU that stores data on which the computer is currently working. Whatever is stored in RAM vanishes when the power is turned off.

- **remote** Not local; happening in a place other than your local machine.

- **ROM** Read-only memory. Permanent memory in the CPU that stores the basic commands the machine needs to boot.

- **root** The very top directory. All the rest of the files are stored in directories and sub-directories within the root directory.

- **RTFM** Read the fine manual that accompanies the device!

- **server** A machine that serves (delivers) data and services to other machines.

- **shell** The command-line interface used by UNIX platforms.

- **software** Computer code that makes the hardware run.

- **UNIX** A popular operating system for workstations and other larger machines. The story behind the name is that once long ago, there was a group working to create "Multics", a do-it-all, or multiplexed, operating system. The resulting OS wasn't very successful, so an effort to create a simpler (!) "uniplexed" system was undertaken by Ken Thompson and Dennis M. Ritchie of AT&T. The OS went by the name of "Unics" which eventually morphed into UNIX.

- **virtual** Not real; happening only on the computer.

- **virus** A program that invades files and software without the user's permission. Viruses may damage files or prevent the computer from operating properly. The most innocuous of the bunch will flash something "cute" or irritating on your screen.

- **wizards** Software (usually within other software) that will execute a task for you. One of the wizards associated with the word-processing program I'm using to write this book keeps offering to help me! I wonder if it knows any physical chemistry?

- **workstation** Usually refers to a computer that is larger and more powerful than a desktop, or personal, computer. However, computer power increases and today's desktop machine is often capable of what yesterday's workstation could do. (This works all the way down the line. The workstation on my desk today has more punch than the supercomputer I used some years back!)

3–3 Programming Languages: Getting a Computer to Do What You Want It to Do

Programming languages are a collection of commands that are essentially shorthand for telling a computer what to do. It may be hard to imagine, but the process of adding two numbers together is a several-step operation for the computer. Rather than having to walk the machine through this process each time, the programmer merely types "+", and a compiler or interpreter translates this into an executable form that includes the more-detailed instructions. Compilers take an entire program (or section of a program) and turn it into executable code; interpreters do the same thing on a line-by-line basis.

The process for writing computer code is more or less the same no matter what language you use. The programmer (you) selects a language and then chooses the desired commands, arranges them in the correct order for the task at hand, asks the computer to turn them into something it can actually execute (using either an interpreter or a compiler), executes the code, and goes home. Right. Well, maybe on a really *good* day. On a regular, run-of-the-mill day, the process might be a bit more circular. You write the code and compile it. The compiler detects some errors. You fix them. You compile it again. It works. You execute. The program crashes. You add some more lines to the code, compile, execute. Crash. Repeat. Finally, the code compiles and executes. The answer to your test problem is wrong. Back to the code again. The longer the code sequence, and the more complex the problem, the more likely you are to repeat the cycle. To give you some idea of how often this will occur, professional programmers are considered productive if they can produce (on average) 10 lines of working, deliverable computer code *per day*.

A plethora of programming languages are available. The following are among those you are most likely to encounter in a physical chemistry class.

- **BASIC** **B**eginner's **A**ll-Purpose **S**ymbolic **I**nstruction **C**ode. BASIC was written by two Dartmouth professors, John Kemeny and Thomas Kurtz, in the early 1960s. It is easy to learn to write short programs in BASIC, but the infrastructure for writing longer programs isn't as nice as in other languages. BASIC is usually interpreted, rather than compiled, and is available for most microcomputers.

- **FORTRAN** **For**mula **Tran**slator. *The* language for scientific and numerical computing. A compiled language. It's not too hard to learn, quite powerful, and generally very fast. The earliest version dates to the late 1950s.

- **Pascal** Named for an eminent seventeenth-century mathematician, Blaise Pascal, this language was designed in the 1970s by Niklaus Wirth. Like BASIC, it is designed to be used by novices, but it has much more structure and many more rules about how different types of variables can be used (that is, it's a strongly typed language).

- **C** Another product of the 1970s, written by Dennis Ritchie, C is a flexible language. Programs written in C are generally easily ported (moved) to other platforms. It's called C because it was derived from an earlier language called B.

- **C++** The successor to C, written by Bjarne Stroustrup of AT&T in the late 1980s. This is an object-oriented language, which means that rather than being organized around "routines" or collections of commands, it's structured around objects. (Think of an object as an overgrown variable. It can include a component that gives the value of something, but it also incorporates the behavior of that something.)

Program Archetypes

The simplest way to write a program to do some task is to modify an existing program. Doing this generally requires some passing familiarity with the language and some fundamental understanding of program structure. Below are illustrated a number of simple program structures. By combining these structures in various ways, you can go a long way toward writing a more-complex program. Obviously, reading this section won't make you an ace programmer, but you should be able to write a simple BASIC program or revise a FORTRAN or C program.

Some Fundamentals

In most languages, the * symbol is used to indicate multiplication, while / corresponds to division and ∧ to exponentiation. Different languages have different rules about the hierarchy of operators, so beware if you are using a new language. Both FORTRAN and BASIC use EPMDAS (Everybody Pleases My Dear Aunt Sally: exponentiation ranks highest, then parentheses, followed by multiplication, division, addition, and subtraction). For example, the expression

$$3 * x^2$$

would be evaluated by first raising *x* to the second power and then multiplying the result by 3. If you wanted the multiplication to happen before the exponentiation, you would use

$$(3*x)^2$$

The expression

$$2*x+12$$

results in the multiplication of *x* by 2 and then the addition of 12 to the resulting product. The expression

$$2*(x+12)$$

gives a different result.

Some languages treat different types of numbers, such as integers and real numbers, differently. Such languages are called "typed". BASIC doesn't distinguish among data types at all. FORTRAN is implicitly typed. It is assumed that variable names starting with the letters i through n are integers, whereas variables starting with other letters are assumed to be real numbers (also called floating-point numbers because the decimal point ("floats"). These implicit requirements can be overridden in various ways. FORTRAN also uses other data types, such as "logical", that must be explicitly declared at the beginning of the program. C is typed more strongly. You must declare each variable and its type at the beginning of the program.

Languages can even keep track of a list of variables, though generally you must reserve enough space to hold the list before you begin. The command that reserves space in BASIC is called DIM (short for dimension). It must come before any other commands.

```
DIM ARR(10)
```

would reserve enough space in a list called **ARR** for 10 entries. Attempting to add an eleventh to the list would result in an error. To get access to one value in the list, use

```
PRINT 1+ARR(4)
```

This would add one to the fourth value in the list and print it out.

In FORTRAN, it works similarly.

```
dimension array(10),index(100)
```

would reserve space for 10 floating-point numbers in a variable list called **array** and for 100 integer values in **index**. Again, you can't exceed the initially declared number of values in any list. If you try, the compiler *may* generate an error message (usually something about array out of bounds). Then again, it may not. You may also create multidimensional arrays in FORTRAN — for example,

```
dimension icolor(256,256,256)
```

This array contains 256^3 elements.

Finally, format matters to some languages. For example, FORTRAN recognizes only commands that start in the seventh column. Columns 1 to 6 are reserved for line numbers and other special flags.

Printing Something

In BASIC this is completely straightforward:

```
PRINT 1.0
```

results in the output of

```
1.0
```

Numerical expressions and variables can also be printed:

```
PRINT (12+4)•(3+2)
PRINT X
```

In FORTRAN it's a little trickier. The easiest way is using "free format." Here the command says to write to unit 6 (the usual output) using free format (indicated by the *):

```
write(6,*) 1.0
```

results in the output of

```
1.000000
```

Text can also be printed using

```
PRINT "Hello there!"
```

in BASIC, which prints everything within the quotes (but not the quotes themselves) or

```
write(6,*) "Hello there!"
```

in FORTRAN.

Both BASIC and FORTRAN allow lists of data and text to be printed using the same command. In BASIC the items on the list are separated by a semicolon:

```
PRINT "When adding 12 to 2 the answer is "; 12+2
```

whereas FORTRAN uses a comma:

```
write(6,*) "When adding 12 to 2 the answer is ",12+2
```

FORTRAN input and output can also be formatted. A formatted output statement in a FORTRAN program would look like

```
       write(6,1000) index,SomeFloat
1000 format(1x,'The ',i5,'th value is',f10.6)
```

The formatting command **1x** indicates leave a single space at the start of the line. The integer value **index** would be printed using at most five digits (**i5**). The floating-point values **SomeFloat** would be printed using ten characters, six after the decimal point. If the value of **index** were larger than 99,999, it would not print; asterisks would appear instead (*********).

Example 3.1 Enter and run the following BASIC program.

```
PRINT "Greetings!"
PRINT "Now I'll do some addition…."
PRINT "12+13 = ";12+13
```

Modify it to print **Greetings and Salutations!** instead of **Greetings!**. Print a different sum.

Reading a Value

Data can be entered into a BASIC program using the **INPUT** statement.

```
PRINT "This program will print the multiples of N."
INPUT "What number would you like?";N
```

This program fragment would do the following:

```
This program will print the multiples of N.
What number would you like?
```

The program will now wait for the user to type a value for **N**. FORTRAN input is very similar to the output. To read a value from a user,

```
read(5,*) ivalue
```

would read from unit 5 (the default input, usually the screen in an interactive program) an integer and store it under the variable name **ivalue**.

Loops

One of the most appealing features of computers is their ability to do repetitive operations. To do a "loop" in BASIC, the following statements are used: **FOR i TO j STEP k** and **NEXT**. Here **k** must be an integer.

```
FOR i=1 TO 10
PRINT i
NEXT i
```

would print the numbers 1 to 10. Everything between the **FOR** and the **NEXT** statement is executed each time the loop is passed through. Loops can be made to run backwards as well.

```
FOR i=1 TO 10 STEP -1
```

Loops in FORTRAN look like

```
        do 100 i=1,10
        write(6,*) i
100 continue
```

Loops can be nested.

```
        do 100 j=1,10
        do 100 i=1,10
        write(6,*) array(i)*array(j)
100 continue
```

Example 3.2 Write a BASIC program to ask the user to enter 10 numbers, sum them, and then print the list, along with the sum.

Tests

Test functions can be used to control the progress of a program. The first part of the command requests a particular condition be tested. If the condition is met, then the rest of the command is executed. Otherwise, it is ignored.

```
DIM ARR(5)
FOR J=1 TO 5
INPUT "Pick any number greater than 10.";ARR(J)
IF ARR(I) <= 10 THEN INPUT "Please try again, the number
must be bigger than 10!"; ARR(J)
NEXT J
PRINT "Your list is:";ARR(1);ARR(2);ARR(3);ARR(4);ARR(5)
```

This code fragment will test each number that the user enters to see whether it is larger than 10. The user is given a second chance to enter the value if it is larger than 10. "If" statements in FORTRAN take the form

```
if (i.gt.10) write(6,*) 'This number is larger than 10!'
```

The logical statement within the parentheses is tested, and if it is true, the rest of the statement is executed. Commonly encountered logical operators in FORTRAN are **.eq.** (equals), **.ne.** (not equals), **.ge.** (greater than or equal to), **.le.** (less than or equal to), **.gt.** (greater than), and **.lt.** (less than). The logical operators **.and.** and **.or.** can also be used.

More-complicated test structures are possible as well. In BASIC,

```
IF x<10 THEN PRINT "x is less than 10" ELSE PRINT "x is
bigger than 10"
```

will print the first comment when **x** is less than 10 and the second when it is greater than 10.

Advice to the Reader

Comments can (and should!) be included in your code. You may not remember exactly what your logic was when you return to a piece of code. If other people want to modify your code, comments can help them see your thought processes as well. The **REM** command in BASIC does nothing, except to **rem**ind you of whatever you type after it.

```
REM       THIS IS A TEST PROGRAM
```

A FORTRAN comment is indicated by putting a C in the first column of the statement line.

```
C         This is a program to compute the average
C         of a list of up to 1000 values
```

Example 3.3 Write a program to find the average of a list of up to 10 numbers. Modify the program to find the standard deviation.

Example 3.4 A more-complicated exercise is to write a program to numerically integrate the function $10.0 - x^2$ over a range from a to b. Use the rectangle rule to do the integration. Modify the program to compare the numerical answer with the answer you get analytically and to compute the percentage error between the two.

Debugging

Not every program works perfectly the first time. In fact, it's probably safe to say that programs of any complexity at all rarely work the first time. There are a few fundamental tricks to figuring out what is wrong with a program.

- **Build up code.** Test drive fragments of code to be sure they do what you want them to do. It's a lot easier to find the error in 10 lines than in 100 lines.

- **Have a test problem.** Be sure you have a problem available to which you know the answer. For more-complicated tasks, you might want to have several test problems that allow you to test all the features of your code. Note that even if all your test problems work, you may still have a bug or two lurking around. Except for very simple systems, it is impossible to run through all possible scenarios.

- **Put in write statements.** You *think* the code is getting to that **FOR/NEXT** loop, but is it really? How many times is it going through? What is the value of the array element before you modify it? Use print statements to print as many checkable intermediate values and locations as you think necessary.

- **Think like the computer.** Work through the code with your test data a line at a time, so you know what values are expected at each point. Check the values with write statements.

Not surprisingly, commercial packages (IDEs or integrated development environments) are available for many computer languages that can assist you in writing and debugging programs. For what you will need to do in most physical chemistry courses, they are probably not worth learning to use if you haven't already mastered them. If you go on to program at more length in a research group, they are certainly worth the effort to learn.

3–4 Getting Down to Business: Using Spreadsheets

Spreadsheets are often associated with business applications, but they can be easily adapted to scientific applications. Their greatest strength is that you can paste in data from other applications. This can be useful when you have a large collection of data acquired by computer. Many commercially spreadsheets are available for a variety of operating systems, ranging from personal digital assistants such as the PalmPilot to Mac and PC platforms. One size never fits all, and the examples that follow are based on the Mac version of Excel. They are meant to illustrate what a spreadsheet can and cannot do for you in a physical chemistry course (besides keep track of your grades, which is what I use it for!). Generally, what one software package can do, its competitor can do too. The trick is to know what can be done and then figure out how to do it using the particular version and platform that you have.

The Basics

Spreadsheets work using "cells" — little boxes that can contain data, descriptive text, or formulas. The cells are identified by a number and letter combination (for example, the cell at the top left is designated A1). You can move from cell to cell horizontally using the tab key; the return key moves you vertically. The arrow keys and mouse can also be used. Values in a cell can be changed by typing over them. If another cell depends on the value, note that the dependent cell won't change until you leave the changed cell.

Arithmetic Manipulations

This is where spreadsheets shine. You can take a set of data from an experiment and easily manipulate it. Take the following set of data, for example.

	A	B
1	temperature in K	rate const
2	700	0.011
3	730	0.035
4	760	0.105
5	790	0.343
6	810	0.789
7	840	2.17
8	910	20
9	1000	145

Cells A1 and B1 contain title information; the first piece of data is in A2. To determine the Arrhenius parameters, you need to plot natural log of the rate constant versus the inverse of the temperature. The easiest way to do this is to use a formula. Title column C (in cell C1) "1/T" and then in cell C2 type

    =1/A1

where the = sign indicates you are typing a formula, not text. This will compute the inverse of whatever is in cell A1. You could repeat this for cells C3 to C9, but there is an easier way. Highlight cell A1 and then copy it. Highlight cells C3 to C9 and paste in the copy of cell A1. *Voilà!* The inverse temperature has been computed.

Computing the natural log of the pressure is nearly as easy. Again, in cell D2 type the following formula

`=LN(B2)`

Using the copy and paste technique of the previous paragraph, you can compute the logarithmic pressure for all the values. Watch carefully when using these built-in functions. For example, the function **LOG** gives you the base-10 log. Other programs (such as Mathematica) use **LOG** for the natural log. Computers are a great way to do many calculations quickly — they are also a great way to repeat the same mistake many times!

Fitting Data

Someone once said that physical chemistry is the art of fitting data to a line. My experiences in physical chemistry haven't done a lot to disprove this aphorism. Let's see how we can use a spreadsheet package to find the best-fit line for the data in the previous section.

	A	B	C	D
1	temperature in K	rate const	1/T	ln(k)
2	700	0.011	0.00142857	-4.50986
3	730	0.035	0.00136986	-3.3524072
4	760	0.105	0.00131579	-2.2537949
5	790	0.343	0.00126582	-1.0700248
6	810	0.789	0.00123457	-0.236989
7	840	2.17	0.00119048	0.77472717
8	910	20	0.0010989	2.99573227
9	1000	145	0.001	4.97673374

To find the slope, type, in a convenient (unused!) cell,

`=SLOPE(D2:D9,C2:C9)`

Note that you specify the *y* data first by giving the starting cell (D2) followed by a colon and then the last data cell; the *x* data is specified the same way. The intercept is equally simple to compute:

`=INTERCEPT(D2:D9,C2:C9)`

If you want to avoid typing the cell values, you can just type the **=INTERCEPT** (then highlight the *y*-values you want, hit the comma, and repeat for the *x*-values. Put on the closing parenthesis and hit return, and you've done it.

Excel allows to constrain your line to run through the origin, if that is necessary. The function for doing this is

`=LINEST(D2:D9,C2:C9,FALSE)`

Plotting Data

Good practice demands that you plot any data you fit to a line (or other function) to be sure that it indeed has the correct form. The chart wizard in Excel can walk you through the procedure. A scatterplot plots just the points. To plot the final line, you need to have points on that line to plot. Generate them using the formula:

```
=-(22650•E2)+27.7
```

One of the disadvantages of Excel is that it is difficult to plot the discrete data as points and the line as a line on the same plot. This is much easier to do in Mathematica. You can plot the line connecting the actual data points and the fit line on the same graph, but this isn't terribly good form.

Bar graphs are easily generated. Again, the wizard is your friend.

3–5 Leave the Solving to Us: Using Symbolic Math Packages

Symbolic math packages are to algebra and calculus what calculators are to arithmetic — a great aid, as long as you know what you want to do. Several packages are widely available, including Mathematica, Maple, and MathCad. The examples below are drawn from Mathematica, but the principles they illustrate hold for any of the major software packages. The examples that follow are not meant to be comprehensive by any means (the current manual that accompanies Mathematica is roughly 1,400 pages long!), but to give you a sense of what can be done that is useful in a physical chemistry class. The best thing to do is sit down and try these functions, using Mathematica or whatever software you have available. Print samples of the procedures that you find most useful so you can refer to them later. Without a reference, it's easy to forget a comma or semicolon somewhere, and a few well-selected examples are much easier to cart around and sort through than the 1,400-page manual.

The Basics

Multiplication in Mathematica is indicated by either the ***** symbol or by a space. To raise to a power use **^** (for example, 2^2 is written **2^2**, and 10^{-34} is written **10^-34**). The natural log function is given by **Log**, the trig functions by **Sin**, **Cos**, **Tan**, **ArcCos**, etc. The value of π is specified using **Pi**, the value of Euler's constant e is given by **E**, and infinity is defined as **Infinity**. The square root of –1, usually designated i, is given by **I**. (The latter is an easy one to get caught by — for example, when using **I** to represent moment of inertia.) If you have an expression involving constants (either constants you've defined or those defined by Mathematica), you can force Mathematica to give you a numerical answer by following your expression with **//N**. For example, to get the numerical value for 2π, use **2 Pi//N**. Previous results can be accessed by using the **%** symbol. For example, the sequence

```
2^2
3^%
```

returns 3^4. The penultimate result is designated **%%** and other results are referred to using **%N**, where **N** is the number of the output cell. Finally, Mathematica is quite particular about input. To evaluate a cell, you must hit the ENTER key, not the RETURN key. **Sin** must be used, not **sin**. Be careful not to confuse braces, brackets, and parentheses. [, {, and (are all used in different contexts.

Solving Equations

To see how equation solving can work for you, consider trying to solve the 2×2 secular determinant

$$\begin{pmatrix} aa - E & b - ES \\ b - ES & ab - E \end{pmatrix}$$

for the energy levels, E. First expand the determinant using

Expand[(aa-E)(ab-E)-(b-E S)^2]

which gives

$$aa\,ab - b^2 - aa\,E - ab\,E + E^2 + 2\,b\,E\,S - E^2\,S^2$$

These energy levels can be found by setting the determinant to zero and then solving for E.

Solve[%==0,e]

The result is

$$E = -\frac{1}{2}\left\{\frac{-aa - ab + 2bS}{1 - S^2} \pm \sqrt{\frac{-aa - ab + 2bS}{1 - S^2} - \frac{4(aa\,ab - b^2)}{1 - S^2}}\right\}$$

You now have an algebraic expression for the energy levels, which you could then plot as a function of b, for example. Note that there is more than one solution; the solutions are given in a list enclosed in { }. You can extract a single solution for use later as follows:

E1/.%15[[1]]

This assumes that the output of the **Solve** command is in output cell number 15. This command will take the first of the roots of the equation and assign it to the variable **E1**. The pesky "{E→..}" will vanish.

Note that Mathematica will not always be able to help you. Sometimes it will be easier to solve, or at least simplify, the equation yourself. For example, consider trying to find the nodes (the points where the wavefunction is zero) for the first excited state wavefunction of the one-dimensional particle in a box.

```
Solve[(2/a)^.5 Sin[2 Pi x/a]==0,x]
```

Mathematica at least warns you that it won't necessarily find all the possible solutions

```
Solve::ifun:
    Warning: Inverse functions are being used by Solve,
so some solutions may not be found.
```

The solution it *does* provide is the trivial solution, zero. Not so helpful.

Evaluating Expressions

You can use Mathematica just like a calculator.

```
3 4
```

returns

```
12
```

Constants can be defined by

```
h=6.62608 10^-34
```

If you'd rather not see the constant (or the result of any other expression) printed, you can use a **;** at the end. Entering

```
h=6.62608 10^-34;
```

will define the constant, but not echo it for you.

You can evaluate an expression by defining the constants and then simply evaluating the expression. For example, the lowest energy level for an electron trapped in a box of length 1Å could be found by using

```
n=1;
m=9.10939 10^-31;
a=10^-10;
h=6.62608 10^-34;
```

and then entering

```
((n^2) (h^2)) / (8 m (a^2))
```

Defining a Function

You can define a function for later use in Mathematica. For example, the "one-dimensional particle in the box" wavefunction for a particle of mass **m** and box length **a** can be defined as

```
psi[x_,n_] := (2/a)^.5 Sin[n Pi x/a]
```

The function can be evaluated at a single point by

```
psi[3,1]
```

If the constants **a** and **m** have been given numerical values, then a numerical value will be returned. Otherwise, the symbols will be used. Common traps here are forgetting to use the **:=** symbol and forgetting to place the _ after the independent variables. Note that you can mix constants (**m** and **a**) with variables (**x** and **n**). Another common error is to use the same name for an independent variable as for a constant that you've already defined. Mathematica allows you to define the same function name more than one way. Then it looks at the number of arguments you provide to decide which one to use. For example, suppose you had two functions **f** defined as follows:

```
f[x_]:=x^2
f[x_,y_]:=x y
```

Then evaluating **f[3]** would give the result **9**, whereas evaluating **f[3,4]** would return **12**.

I don't recommend using this "overloading" technique, at least at first. If you've defined a function already and want to substitute a new definition, you can use the **Clear** function to delete the original definition — for instance,

```
Clear[psi]
```

Integrating and Differentiating

One can find definite and indefinite integrals, as well as numerically integrate functions. For example, to normalize the "particle in the box" wavefunction, one must evaluate the definite integral

$$\int_0^a \sin^2 \frac{n\pi x}{a} dx$$

Use

```
Integrate[(Sin[n Pi x/a])^2,{x,0,a}]
```

Similarly, one can find the normalization constant for the ground-state harmonic oscillator wavefunction

$$\psi_0 = c_0 e^{-\alpha x^2 / 2}$$

using

```
Integrate[(E^(-alpha x^2/2))^2, {x,Infinity,Infinity}]
```

Numerical integration proceeds similarly. The definite integral for the "particle in the box" problem could just as easily be done numerically:

```
n=2;
a=4;
NIntegrate[(Sin[n Pi x/a])^2,{x,0,a}]
```

Note that although one could use a symbolic value (**a**) for the integration limits when computing the definite integral analytically, one must use a numerical value for the numerical integration.

Differentiating is similarly straightforward. If one wishes to find an expression for the enthalpy of a free rotor, one must compute the derivative of the partition function with respect to temperature.

$$H = RT^2 \left(\frac{1}{q_{tors}} \right) \left(\frac{\partial q_{tors}}{\partial T} \right)$$

where the partition function is of the form

$$q_{tors} = \frac{1}{m} \left(\frac{8\pi^3 IkT}{h^2} \right)^{1/2}$$

The appropriate derivative can be computed as follows

```
D[(1/m) Sqrt[8 Pi^3 Iner k T/(h^2)],T]
```

Matrices and Vectors

Mathematica and similar programs are particularly handy for dealing with matrices and vectors. A vector is defined as

```
Xaxis={1,0,0}
```

and a matrix can be defined as

```
SecularMat={{aa-E,b-E s},{ab-E,b-E s}}
```

It can be easier to tell whether a matrix is correct if you can see it printed in the usual form. Use

```
MatrixForm[SecularMat]
```

To take the dot product of two vectors, the **.** is used, as in

```
{1,3,2}.{4,3,1}
```

or

`Xaxis.{1,3,2}`

Two matrices can be multiplied similarly. For example use

`A.B`

to multiple two previously defined matrices **A** and **B**. In both instances, be careful not to confuse scalar multiplication (indicated by a space or *) with the vector and matrix operations. For example, `{1,3,2}.{4,3,1}` returns **15**, but if you try `{1,3,2}*{4,3,1}` you will end up with another vector, `{4,9,2}`, whose components are the products of the components of the original vectors.

You can add vectors — `{1,3,2}+{4,3,1}` yields `{5,6,3}` — or multiply them by a constant — `a {1,2,3}` produces `{a, 2 a, 3 a}`. You can also apply a function to all the components of a vector using **Map**.

`Map[Cos,{1,3,2} Pi]`

evaluates as

`{Cos[Pi],Cos[3 Pi],Cos[2 Pi]}`

or

`{-1,-1,1}`

Taking the determinant of a matrix is trivial

`Det[SecularMat]`

as is finding its eigenvalues,

`Eigenvalues[A]`

To invert a square matrix, use

`Inverse[A]`

The pseudoinverse of a rectangular matrix can be found by using

`PsuedoInverse[A]`

A matrix can be transposed by using

`Transpose[A]`

Repeating Tasks and Summations

Often one wishes to sum a collection of data, or to generate a table of data given a function. These types of programs are ideal for this. For example, to make a table of the relative populations of the first ten vibrational states of HCl (assuming that the molecule behaves as a harmonic oscillator!) I first set up all the necessary constants. Then I create the partition function (**qHO**) and two functions that describe the population distribution (**n** and **PinI**). The **Table** command is then used to create a list (in the form of a vector) of the populations in each of the first 10 states at 298 K.

```
h=6.626 10^-34;
k=1.381 10^-23;
c=2.998 10^10;
wavenum=559.71;
nu=c wavenum
qHO[t_]:= 1/(1-Exp[-h nu / (k t)])
n[i_,t_]:=Exp[-i h nu / (k t)]
PinI[i_,t_]:=n[i,t]/qHO[t]
Table[PinI[i,298],{i,0,10}]
```

The **TableForm** command will take the list generated by **Table** and make a vertical list that is easier to read than the bracketed vector form.

```
TableForm[Table[PinI[i,298],{i,0,10}]]
```

You can create multidimensional lists, as well. The following will generate a list of vectors, each of which contains the state number (**i**) and the value of the percentage of the population in that state (**PinI**). **TableForm** will now make a nice table with the state number, followed by the percentage value.

```
TableForm[Table[{i,PinI[i,298]},{i,0,10}]]
```

If you wanted to sum the values (to show that they come quite close to 1, indicating that most of the molecules are in one of the first 10 states) instead of just looking at them using **Table**, try

```
Sum[PinI[i,298],{i,0,10}]
```

Plotting

Plotting a Function

Using the **Plot** command, you can construct everything from elementary to very complex graphs. A good tactic for any given plotting task is to start simple and add levels of complexity as you go. For example, you could plot the first-order decay of reactant A

$$A_o e^{-k_1 t}$$

using

```
Plot[Ao Exp[-k1 t],{t,0,10}]
```

If you haven't previously defined **Ao** and **k1**, you will get an error message with the note that the function is "not a machine sized real number at". You might at first think that your function values are too large or small at some point, which could possibly be the case, but this error message is usually caused by failure to define some constant or function used in the plot.

Mathematica picks what it believes a reasonable range over which to graph the function, but you can override this and see whatever section of the plot you wish. Use

```
Plot[Ao Exp[-k1 t],{t,0,10},PlotRange->{0,8}]
```

to plot the exponential decay of **[A]** over the range 0 to 8 time units or (if you've already plotted this in output cell **N** and just want to change the range)

```
Show[%N,PlotRange->{0,8}]
```

The display of both the abscissa and ordinate can be controlled using **PlotRange**; for example

```
Show[%N,PlotRange->{{0,8},{2,4}}]
```

will display values for **t** from 0 to 8 and for the **[A]** from 2 to 4.

Multiple functions can be displayed on the same plot. The command below will plot the reactant, intermediate, and product concentrations as a function of time for the following mechanism:

$$A \xrightarrow{k_1} B$$

$$B \xrightarrow{k_2} product$$

```
A[t_]:=Ao Exp[-k1 t]
B[t_]:=(k1/(k2-k1)) (Exp[-k1 t] - Exp[-k2 t]) Ao
Pr[t_]:=Ao-B[t]-A[t]
Plot[{A[t],B[t],Pr[t]},{t,0,10}]
```

You can also display two (or more) previously constructed graphs on the same plot by using the **Show** command. This is particularly useful when you've made one plot using the **Plot** command and another by plotting a set of individual points.

```
Show[%N,%M,%Q,%P]
```

Adding Color

It can be difficult to distinguish among different lines on a plot, and color can help. The percentage populations of the first three vibrational states of HCl can be plotted as red, green, and blue lines, respectively, by

```
Plot[{PinI[0,t],PinI[1,t],PinI[2,t]},
  {t,298,5000},PlotStyle>{{Hue[0]},{Hue[.3]},{Hue[.6]}}]
```

Trying different colors can be a definite time sink, so watch out when using these types of commands.

Plotting Data

Often in physical chemistry you wish to plot a data set. Some tricks exist to make this easier. Consider a problem where you must use vapor-pressure data to find the enthalpy of vaporization of a substance. You are given the following data table.

T in °C	57.4	100.4	133.0	157.3	203.5	227.5
Pressure	1.00	10.0	40.0	100	400	760

The enthalpy of vaporization can be found by plotting the natural log of the pressure versus the inverse of the temperature in kelvins. Mathematica can make the manipulations much easier. First enter "vectors" corresponding to the raw data:

```
temp={57.4,100.4,133.0,157.3,203.5,227.5}
press={1.00,10.0,40.0,100,400,760}
```

Now you can easily convert the temperatures to K and invert them

```
x=1/(temp+273.15)
y=Log[press]
```

The **ListPlot** command needs a list of points, in the format **{{x1,y1},{x2,y2}…}**. This can be constructed from the raw-data vectors as follows:

```
data=Transpose[{x,y}]
```

The data can now be plotted as points

```
ListPlot[data]
```

To plot the data connected by line segments (generally a poor way to display a plot, but you may have your reasons!)

```
ListPlot[data,PlotJoined->True]
```

Plotting in Three Dimensions

Consider plotting the wavefunction for a particle in a two-dimensional box

$$\psi(x,y) = \left(\frac{2}{a}\right)^{\frac{1}{2}}\left(\frac{2}{b}\right)^{\frac{1}{2}} \sin\frac{n\pi x}{a}\sin\frac{m\pi y}{b}$$

A contour plot could be used to display the wavefunction of the first excited state, for example

```
ContourPlot[Psi[x,y],{x,0,a},{y,0,b}]
```

The resulting plot looks like

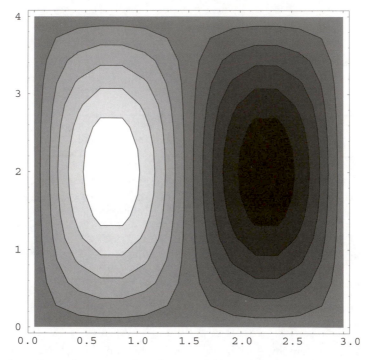

Alternatively, a three-dimensional surface could be viewed by using

Plot3D[Psi[x,y],{x,0,a},{y,0,b}]

with the result

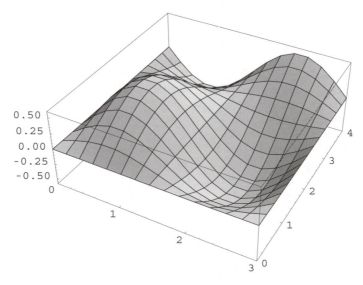

Bar Graphs

Bar graphs can be useful. For example, the data table printed for the first ten vibrational states of HCl might be better displayed as a bar graph. To use this function in Mathematica, you must first load the additional graphics functionalities using

```
<<Graphics`Graphics`
```

The bar graph can be generated using

```
states=Table[PinI[i,298],{i,0,10}];
BarChart[states]
```

Fitting Data

This is another very common task in physical chemistry. To illustrate how Mathematica might be used to do this, let's continue working the problem started above, using the vapor pressure at various temperatures to determine enthalpy of vaporization. To get the slope of the line, use a linear fit

```
Fit[data,{1,x},x]
```

This returns an equation for the best-fit line. You can then plot the line using **Plot**. It is possible to view the line plot and the plot of the points simultaneously (something you should make a habit of doing to be sure the fit is appropriate) by using the **Show** command as described above. Higher-order fits can also be done. For example, a cubic fit can be done by using

```
Fit[data,{1,x,x^2,x^3},x]
```

3–6 Computers in the Laboratory: Data Acquisition

When I took p-chem lab twenty years ago, virtually nothing we did in lab involved a computer. The high point of the quarter was analyzing milk samples for heavy metals via neutron-activation analysis. We were ushered into a basement room and permitted to type a few numbers into a computer that controlled the research nuclear reactor. Now my students use a computer for nearly every experiment, either to control the apparatus or to acquire the data or both (*and* we don't keep a nuclear reactor 50 feet under the p-chem lecture room!). Most instruments used in students labs as well as in research are computer controlled, including FT-IRs, NMR, and GCs. Although the manuals for such instruments can be overwhelming to the novice (and to the expert as well), most departments and instructors provide "cheat sheets" that can be used for common operations.

Many multipurpose computer-interface packages are available, such as LabView and SensorNet. These allow you to control the flow of data to a computer and its display. In a pinch, you can write your own interface program using BASIC to read from the serial port. It sounds intimidating, but it's not much harder than reading from a file. As with many things, seeing an example helps a lot. If you need to write your own code for an interface, check the manual for a sample program for the instrument you are interfacing or one for a similar apparatus. For example, the Spec21D manual includes a short interface program in BASIC that can easily be modified to work on a variety of computer platforms.

3-7 More Information

- *UNIX for the Impatient* by Paul W. Abrahams and Bruce R. Larson (Reading: Addison-Wesley, 1996). This book is a good basic reference for using UNIX.

- *Scientific and Engineering C++* by John J. Barton and Lee R. Nackman (Reading: Addison-Wesley, 1994). Recommended by my brother, a computer scientist at Lawrence Livermore Labs and a C++ guru as a good resource for learning the language, this is best read by those already proficient in another language.

- *Computing for Engineers and Scientists With Fortran 77* by Daniel D. McCracken and William Salmon (New York: John Wiley & Sons, 1988). This book is very good to use to learn FORTRAN. My graduate students have used this one for years, but it is still in print.

- *Basic Programming for Chemists: An Introduction* by Peter C. Jurs (New York: John Wiley & Sons, 1987). This is just what it says — BASIC programming with a chemical slant. A collection of about 50 examples is quite helpful.

- *The Mathematica Book* by Stephen Wolfram (New York: Cambridge University Press, 1999). All 1,400 pages of the fourth edition have some useful piece of information on them. This is a terrific reference with lots of good examples, but it is not easy to cart around. If it's not sitting next to the computer on which you are using Mathematica, lobby for it. It is also now integrated into the Mathematica interface on many platforms — a real relief given its mass.

Chapter 4

The Write Stuff

Recently I spent a semester teaching a "non-science" seminar course for first-year students. When I told one of my students that I thought she wrote very well, she replied, "That's too bad, since I want to be a scientist!" I spent much of the remaining portion of her conference trying to convince her that writing is as critical for a science student as it is for an English major. I don't think I succeeded. For many chemistry majors, an advanced lab course is all it takes to convince them that writing is as necessary a tool for the scientist as her ability to integrate.

Scientists must write and write well for some obvious reasons. First, they need to keep a lucid, contemporary record of the work being done in the laboratory. Next, results must be clearly communicated to colleagues and other interested parties. Finally, groups that support research with grants generally require something effectively written to convince them to part with funds.

Scientists also have some less-evident motives for writing. Writing can clarify one's thinking on a problem. For example, Einstein and Schrödinger exchanged a delightful set of letters as the two tried to thrash out various ideas in quantum mechanics[8]. This correspondence was not meant to be a formal report on Schrödinger's work, but rather served to help both Einstein and Schrödinger sharpen their thinking.

Although this chapter serves primarily as guide to the sorts of writing you are certain to do in a physical chemistry course, keep in mind that the most useful writing you do in the course might not be in response to a class assignment.

4–1 Keeping a Lab Notebook

Often, the ground rules for keeping a lab notebook seem arcane and, frankly, like overkill. "Why use ink when I can use the pencil I've got right here? Plus I can erase my mistakes that way." "I'll hand my lab notebook in when I've copied it over. You couldn't possibly read it now anyway!" "Oh, *those* spectra. Must be somewhere in my room." Every instructor and employer will have specific (and idiosyncratic) guidelines for keeping a record of your laboratory work. I've tried here to give you some advice that will apply in most situations, as well as some of the reasoning behind the odd "rules".

---

[8] *Letters on Wave Mechanics: Schrödinger, Planck, Einstein, Lorentz,* edited by K. Przibram for the Austrian Academy of Sciences, translated and with an introduction by Martin J. Klein (New York: Philosophical Library, 1967).

The Whys and Whats of Notebooks

What is a lab notebook? It is a complete, contemporary document of laboratory work. Work is recorded as it is done. The record not only includes observations and results of experiments but should also provide insight into the thought processes behind the work. Thus, instead of merely recording the results of experiment X (let's say the an NMR spectra of a previously synthesized compound), a good notebook would record what led you to do X ("I think I can confirm the structure of compound **3** by NMR. Two possible structures, suggested by the IRs on pg. 23, are…."), along with an analysis of the results.

Why do such a thing? Probably the principal reason why you are asked to keep a lab notebook in advanced lab courses is to train you to keep a research notebook. The reasons for keeping a good research record are myriad. First, it enables you to keep track of a project. Trust me: after a point, you can't remember it all. Most research projects take months or years to complete and you will have a tough time recalling exactly why you ran the spectra taped to page 47. Even in physical chemistry courses, where the work generally extends over just a couple of weeks, you can forget exactly what the settings were on the instrument when you took your data. Second, your research notebook provides a reference tool when you want to collect results to share with others. Third, it allows someone to continue your work without having to consult you directly. It's generally easier for a research advisor to locate a former student's notebook than the former student. In the case of employers, access to research notebooks can minimize the impact when employees quit or die. Finally, the research notebook is a historical and legal document. It can be proof of what was found when. In arguments over patent rights or in the case of disputed results[9], the original notebooks can be cited as evidence. And last but not least, if you become famous, it will give history-of-science Ph.D. students something about which to write a thesis![10]

How?

- Use a bound notebook — not a spiral-bound notebook, and not a three-ring binder. Some books have carbon pages that you must turn in to the instructor. The point of using a bound book is that you can tell when pages have been removed (necessary if the notebook is going to be used as evidence). More practically, pages fall out of three-ring binders, particularly when they are referred to frequently.

- Use ink, preferably not water-soluble ink. The reason for the latter should be obvious to anyone who has been in a wet lab. The purpose of using ink at all is that it can't be erased. Again, this preserves the record in case of dispute. Most of us will never have to produce our notebooks for this purpose. But if you erase it, it is still gone, and you just might discover that the number was right after all!

- Date and number your pages (this makes referring to them easier).

---

[9] Cases of scientific fraud come to mind here.

[10] For a fascinating look into Pasteur's notebooks, see *The Private Science of Louis Pasteur,* by Gerald L. Geison (Princeton, New Jersey: Princeton University Press, 1995).

- Write as you go. This is meant to be a contemporary record, which means that entries should be written as soon as possible after the actual event. Lab notebooks shouldn't be updated once a week, and they should never, ever be recopied. If your handwriting is awful, you just need to learn to write as clearly as you can for this purpose, and those who come after you will just have to cope with that effort.

- Be verbose. Record as much as you can, even if at the time it seems excessive. It can be quite frustrating to discover you need the temperature of a water bath, which at the time didn't seem critical. Try to think ahead about what might be reasonable to record — it can be harder to do this in the heat of the moment. Often you can establish a list of common things that must be recorded each time you work. Things that might usefully be recorded include temperature and humidity of the lab, air pressure, sources of chemicals, instrument types, and instrument settings. Remember to sketch your thought processes as you go, too.

- Draw. Sketches are worth a great deal. Sketch the glassware setup you used, the structure of the molecule just calculated, the instrument setup, the appearance of crystals, and so on.

- Label. Don't just write down a number; give a brief description. For example, instead of "0.1279", enter "0.1279 g of NaI weighed to prepare a 0.015 M solution of iodide".

- Everything that can practically go in, should go in. Unless you are forbidden to do so, tape or staple spectra, programs, spreadsheets, etc. into the lab notebook. Things that are too large to put in reasonably should be organized somewhere, and their location should be recorded in the notebook. For example, my 10,000 lines of computer code are obviously too much to put in (I generally never even print the whole thing). I *do* note in the notebook where I can find the code ("/mfrancl/pchem/kinetics/analysis.for on kelvin") and where it is backed up ("backed up on 2/13/99 to Zip").

- Library work belongs here, too. References to procedures, literature values, and so on should be entered fully at least once. Don't just give the literature value; record where you found it, too.

- Be exacting about units and abbreviations. Always specify the former and keep track of the latter somewhere. Brevity is great, but if you can't remember what Fg stands for, you can be in for a rough time.

- Mistakes should be crossed out. Don't obliterate them entirely (no matter how tempting!); you may want to be able to read them later. I generally write a short note explaining why I think the entry is incorrect and refer to a later page with a fuller explanation.

Finding Things

Keeping a contemporary record means putting up with some disorganization. The entropy calculated last week on page 67 needs to be compared with one done today. In between, several structures were calculated and some literature values located. Librarians maintain that you don't need a catalog of a book collection until it exceeds about 10,000 volumes. Similarly, if you have just one or two notebooks, you can probably recall roughly where most of your critical data is. Some tricks, however, can help you keep track of things in a lab notebook. All of them require some extra investment of time and may or may not be worth it, depending on the length of your notebook and the number of times you expect to refer to the information.

You can create an index of structures, spectra, and/or compounds. I used one of these in graduate school to keep track of all the molecular structures I had generated. I kept it on the computer and added to it weekly. I could then print a current version sorted by empirical structure whenever I needed it (for example, when writing my thesis).

A table of contents can also be effective. For each page, write a one- or two-line description of the contents ("119 structure of $Ph_2CO/AlMe_3$"). These descriptions are easy to skim when you look for information. If you begin the contents on the last page of the notebook, rather than on the first page, and put subsequent contents pages in reverse order in front of the first you can avoid running out of room. In other words, the first page of the table of contents is placed on page 100 of a 100-page notebook, the second page of the table of contents is placed on page 99, and so on. It can be frustrating to have reserved three pages in the beginning for a table of contents, and in the end need five pages. No matter what the type of your research or how you structure a notebook, it is probably a worthwhile habit to keep a table of contents.

Figure 1
This page from the laboratory notebook of James Mason Crafts (of the Friedel-Crafts reaction) illustrates many of the desirable features of a record of experimental work. The page has been numbered and dated (the notebook is from 1877; Crafts doesn't record the year on this page). An error has been crossed out, the apparatus is sketched, units are given. Crafts notes "NB the temperature was taken in the ag bottle". At the end, he reports "this was too little to purify".

Periodically collect your results from within the notebook. I recently collected all the absolute entropies I'd calculated over the last six months into a large table. Each entry is cross-referenced to its location in the notebook, which makes it easy for me to go back and check on the details.

Sticky notes are often useful in temporarily marking critical pages in the notebook. The danger here is that you will write something on them. If it isn't firmly attached to the notebook, it shouldn't have anything written on it!

To sum up, a well-kept lab notebook is a tremendous resource, both for you and for those you work with. Clarity and completeness count for a great deal, neatness for far less. A good lab notebook should have that lived-in look!

4–2 Lab Reports

Good writing is good writing, no matter whether it is for a science course or an English course. Writing that excites and draws the reader into the work will always be appreciated, whether the work is a novel or a physical chemistry lab report. A good lab report will also be, as one teacher of science writing[11] would put it, "politically appropriate". In other words, the writing is tailored to a specific audience and has a particular goal or goals to be achieved. The key is knowing what approach to take to meet the desired goal with the given audience. For physical chemistry, the goal is to convince your instructor that you correctly executed the laboratory work, that the results of that work are correct to within some limit, and — most critically — that you understand the significance of your results.

Although every writing project is a little bit different, I apply two basic principles across the board: (1) my writing can always improve, and (2) organization and discipline are critical. The suggestions that follow draw from these principles, along with my more than two decades of experience writing in science and fifteen years of reading p-chem lab reports!

Getting Started

This is by far the hardest part of any writing project! Begin by reviewing the obvious constraints. Know the due date. What is the policy on drafts? Are they optional? Required? Be clear about the format and content your instructor requires. Neither reorganizing your lab report nor scrambling to find some piece of experimental data from the literature is going to be a pleasant experience at the last minute.

Once the constraints are established, take a moment to make a plan. I find a checklist to be helpful. I get a sense of relief when I check off a completed piece, and doing so also helps me to not miss any critical parts. Many journals and granting agencies include checklists for their authors — you can easily construct your own. Be sure to give yourself deadlines for each part!

---

[11] *The Craft of Scientific Writing* by Michael Alley (New York: Springer-Verlag, 1996).

Write Early and Often

You don't need to wait for all the lab work to be completed to start writing. If you begin early, you will not only be less stressed when the lab is due, but also have a better report. Furthermore, writing more frequently, for shorter periods, lessens the burden of an extended writing project.

Starting to write while the lab work is still going on allows you to clarify your thinking. For example, make it a practice to write a draft of the introduction or purpose for a lab after completing the first day's work. Later work in the lab will benefit from the clearer understanding you will have of the material. Also work up a sample calculation, even if you don't have all the data. If necessary, make estimates or even guess. This can help point out holes in your record keeping (yes, you really did need to take the temperature of the bath), as well as increasing your comprehension of the material. Start tracking down any literature material you might need — for example, previously measured values or references to procedures.

The ultimate benefit of having words and numbers on paper early is the opportunity to review it. Nothing improves writing, including science writing, like editing.

Mechanics Matter

At the very least, grammatical errors in a report are distracting and annoying to the reader. In the worst case, they can obscure the scientific content of your reports. Though your reader is likely to be merely annoyed if you confuse *it's* and *its,* a run-on sentence or a sentence fragment constitutes a bigger obstacle to clarity. Although the mechanics matter, this isn't the place to teach you where to put the commas or when to use a semicolon. If you need help with these tasks, consult a good writing manual (several recommendations appear at the end of the chapter). Many colleges and universities also have writing centers at which you can have a third party read and comment on your work. Word-processing programs often have grammar assistance built in, though not all the advice may apply to science writing.[12]

When I entered high school, my father advised me strongly to learn how to type. He was convinced that his skill at typing had earned him his chemistry Ph.D. more quickly than it otherwise would have been conferred. He suspects that his advisor, confronted simultaneously with his typed draft and his lab mate's draft (a box of handwritten notes on napkins and scratch paper), read his first. I'll update his advice. Learn to use a word-processing program effectively. This is best done before you are confronted with a lab. Below, I've highlighted some of the word-processing features you are most likely to find useful in preparing a lab report.

- **Use the spellcheck feature.** Even though this is not foolproof and is certainly not a replacement for proofreading, it is your first line of defense. Spelling errors suggest a lack of attention to detail, which leads the instructor to wonder whether you fail to pay enough attention to detail in the lab as well! Be careful about scientific

---

[12] One of the versions of Microsoft Word that I used reminded continually me that *orbital* was an archaic usage and I should consider alternatives. As you might imagine, given that molecular orbital theory is my field, this comment came up frequently!

terms. I read one student report in which *ketone* was misspelled as *keytone* throughout. If you aren't certain how a term is spelled, check! You can often teach your word processor to recognize words. For example, in Microsoft Word you can do this by choosing to add an unrecognized word to the standard dictionary.

- **Figure out how to do subscripts and superscripts.** Yes, you can always write them in by hand, but you will have to do so each time you print the material, thus you risk forgetting to put them in some version. I recently graded a student's report where I couldn't tell the difference between I_2 and I^- — a critical distinction — because she had forgotten to write in the subscripts. The same thing applies to Greek letters and to symbols such as $\pm$ or $\approx$.

- **Learn to produce a simple table.** This is definitely a worthwhile exercise. Some word-processing software will even automatically format material pasted in from a spreadsheet program.

Two advanced skills that might be worth an investment of time if you intend to take more lab courses or to write a thesis are embedding figures into your document and embedding equations. Some word processors come with features that allow you to produce equations; others require an additional piece of software. Many physicists and mathematicians are enamored of a program called Tex, which is designed for producing documents that contain a large number of equations.

A Sample Format

The following is a formal description of a fictional experiment. The format is the one used in my course and is based on that used in the chemical literature. Interspersed throughout the report are comments in italics that further clarify the contents and format of each section. Your instructor will undoubtedly specify a different format, but the general features are not likely to be very different.

The Preparation and Analysis of a Chocolate Dessert: Brownies

The title of the report should be brief but informative. The title should describe the work and allow the reader to determine whether the report is of interest. "Spectroscopy Lab" is not an adequate title.

Abstract: Following literature procedures, 24 chocolate brownies were prepared using a conventional oven. The products were analyzed using an in vivo assay. A yield of 40% was obtained under the experimental conditions.

*The abstract should **briefly** describe the experimental procedure followed, the system studied, and the **final outcome** of the work. If a single value or a small collection of values was obtained, include the numerical results in the abstract. The abstract should give the reader all the information obtained from the experiment (or group or experiments). The abstract is **not** a "teaser" to tempt the reader to read the rest of the report — it's a summary. **The reader should be able to understand the main points by reading the abstract, exam-***

ining the tables and figures (with their detailed captions) and reading the conclusions.
An abstract is typically fewer than 300 words.

Introduction: One of the earliest recorded preparations of a chocolate confection was by Montezuma *et al*. The resulting drink, while satisfying, was not easily portable. The synthesis of a more portable, solid product is thus desirable. The cacahuatl drink is a homogeneous liquid, in contrast to the brownies prepared by Boynton, which, though solid and more easily handled, include chunks of pecan. Modifying the procedure of Boynton, we have succeeded in preparing a solid, homogeneous chocolate confection.

The introduction should inform the reader of the significance and objectives of your work and provide the appropriate background. Unless offering a comprehensive review of the literature is the purpose of the report, include only those references that are directly relevant to the work at hand. Do not put the procedure into the introduction. Give the structures of relevant compounds. Relevant equations should be placed in the text. Give highlights of the derivation of major equations. Define any abbreviations used, unless they are in broad usage (for example, NMR is fine, but PMSE should be defined). The length of the introduction depends on the nature of the experiment.

Procedure: The brownies were prepared following the procedure described by Boynton in *Chocolate: The Consuming Passion*. The procedure was modified to eliminate the pecan halves. A 25-cm square glass pan made by Corning was used. The brownies were baked in a conventional oven. Bittersweet chocolate was obtained as chips from Nestlé and used as supplied. The purity was checked by an in vivo assay of samples drawn from the stock supply. All-purpose flour was obtained from Pillsbury and sifted prior to use.

Brownies were tested for doneness by sampling one square of brownie mix. When crumbs resulted, the brownies were considered "done".

*In a description of the procedure, you should **not** repeat instructions easily obtainable elsewhere. Careful note should be made of all differences between the procedure as given in the literature and the one that you followed. If the procedure is new, you should provide sufficient detail for another **chemist** to reproduce it. In other words, don't bother giving detailed instructions on how to rinse a pipette, but do describe any specialized glassware or equipment used.*

*In addition to a stepwise procedure, you should also include such information as the model and manufacturer of **critical** pieces of equipment, because the performance of the instruments may affect the precision of the results. For example, the frequency of the NMR spectrometer used could be important to someone trying to repeat the work, whereas the model number of the analytical balance used is not likely to affect the results.*

The sources of chemicals used can also be important. Different methods of manufacture can leave in the reagents different impurities that could also affect the results obtained. If the chemicals were not used as supplied, described how they were prepared prior to use, such as by distillation. The purity of the materials is also noted, as well as the method by which purity was determined.

Calculations:

Theoretical yield of 5-cm² brownies per batch:
  25 cm × 25 cm = 625 cm²/batch
  625 cm²/5 cm² * brownie⁻¹ = 25 brownies   (theoretical yield)

Actual yield was determined by hand count as 10.
  10 brownies/25 brownies × 100% = 40%   (actual yield)

In this section, you should lay out the method by which numerical results were obtained. If a computer program was used, this should be noted and, if appropriate, included as an appendix to the report. The formula used to calculate each number should be given, and for complicated calculations, a sample should be worked. It is not necessary to type this section. Neatly handwritten calculations are generally acceptable and much less time consuming to produce.

Results: A yield of 40% was obtained. The results of the analysis (Table 1) showed that all testers found the brownies to be acceptably chocolatey and easily handled.

Table 1

Sample Size	Chocolatey?	Ease of Handling
4	4	4
6	6	6

The Results section should contain just that — your results. It is usually best to present the results first and then discuss the possible errors and significance in a separate section. Tabular or graphical presentation of data is nearly always clearest. When using graphs, you can include more than one set of data on a single sheet. Be sure to use different styles of lines or points to differentiate the data sets.

All tables and graphs should have captions. Give each table or figure a number for easy reference from the text. A reader should be able to understand the main points of the report by reading the abstract and examining the tables and figures (including their captions).

Discussion: The preparation of a solid, homogeneous chocolate confection was successful. However, the yield was found to be very low because of the destructive sampling procedure. A higher yield could be obtained by sampling of the reaction mixture less frequently, or by using a non-destructive sampling method, such as the fork method advocated by Boynton or a spectroscopic method (are they brown on top or black?). The baking time was found to be about 10% longer than that reported by Boynton, as a result of either the glass pan or the lack of pecans. Because the type of pan used was not given by the previous workers, further work will be required to establish the cause of the extended baking time.

This is really the most important part of the report. Here you should describe the significance of your results and discuss ways to improve or extend the work. The length of this section will vary considerably, depending on the type of experiment. Did the experiment work overall? Comment on the percent error in numerical values, describe any systematic errors that you believe occurred, and suggest improvements to the procedures that might overcome these difficulties. Indicate what other experiments related to this work could be done, and answer questions that these results raise.

Conclusions: A good recipe for chocolate brownies was presented. The yield is low but could be improved. Further work is in progress.

This section is really just a recap of the abstract. Repeat any important data or constants that were found as part of the experiment. If more work is planned, this is usually noted here. When a report has been particularly long and has presented a lot of results, this section is a valuable tool for your reader.

Acknowledgments: The author would like to thank those who sampled the brownies.

It is customary to acknowledge any financial support for the work performed, as well as the help of colleagues and/or support facilities. For example, a helpful series of discussions with someone or the assistance of the computing center might be noted.

References:

List all the books and articles you used in the course of doing the work and preparing the report. Referencing styles differ among disciplines and even among journals in the same discipline.

Appendices:

In the chemical literature, supporting data, long derivations, and short computer programs are often included as appendices.

Getting — and Using — Feedback

Good writing has been reviewed and edited. The reviewing process can be either formal or informal, done by the writer or by someone else, and done once or several times. The key is always *using* the comments generated by whatever review process you choose.

What sorts of things should a reviewer look for? A reviewer should obviously be alert for typographical and formatting errors. A few of the questions given to reviewers for the journal *Organometallics* by its editor, Prof. Dietmar Seyferth, offer a good starting point:

1. Please give special attention to the experimental section.

- Is the description of the experiments sufficiently detailed so that a reader could repeat them?

- Are new compounds adequately characterized?

- Is all the chemistry reported in the results and discussion sections adequately backed up by experiment?

2. We have an annual page budget, so journal space is valuable. Please read the manuscript with this in mind.

   - Is the introduction overly long? Could the results and discussion sections be made more concise?

   - Is the description of experiments overly repetitive? Are there redundancies in data presentation in the experimental, results, and discussion sections?

   - Can some of the more routine data and some of the figures be transferred to the Supporting Information package?

3. Are the title and abstract sufficiently informative? Are abbreviations defined?

4. Is there prior relevant literature that should have been cited and discussed?

Nearly everything here applies to a good physical chemistry lab report as well as it applies to a paper acceptable for publication in the journal. Even though you don't have a page budget, your instructor has a lot of reading to do. Don't feel obliged to fill space on the theory that longer is better — it's not. To this list, I would add a final question specific to lab reports: Is it clear from the text that the student understands the principles behind the lab?

The simplest review is self-review. Read what you have written, and do so more than once. Keep in mind the specific questions above. It can be particularly helpful to read your work aloud. This slows you down, making it more likely that you will find errors. A reading done a day or so after you have finished writing will also help you identify areas that seemed clear as you were writing but, now that your recollection of your thought processes has dimmed, are less clear. A reader not privy to your original thoughts is even more likely to find these sections obscure.

Often you can exchange reports with someone else in the course, or even find a student who has already taken the course. Give these readers a specific list of things you'd like them to pay attention to when they read the document — for example, "Is my introduction complete?" and "Is the derivation of equation 10 clear?"

The gold standard of reviews is that of your instructor. Many instructors will read drafts, either on a voluntary or a required basis. If the review is voluntary, take advantage of it. This will serve two purpose. The most obvious outcome is that you will get information on how to get a better grade. Less obviously, it will keep you from procrastinating, which will also result in a better grade! If *this* reviewer isn't happy, be *sure* to make the changes asked for.

Once you get back comments from a reviewer, be sure to address them in a subsequent draft. When I read the final copy of a report, I often see that students have changed the typos I marked and may also have changed words that I indicated were inappropriate, but it is rare that they address more-substantive comments. (For example, I might have commented that the draft abstract was a "teaser" for the paper, rather than a summary. I expected the student to rewrite the abstract to better present the results of the experiment.) If a reader is confused by a section, try to clarify it, even if this means adding material. In short, respond to the comments of your reader as you might if the two of you were having an oral discussion of the report.

4–3 Searching the Literature

A phone call from our science librarian was the first clue — I'd lost my class. "A number of your students have been up here trying to find a literature value for the combustion of napthalene. They don't have any leading references. Do you?" "It's actually in the back of their textbook," I replied, "They shouldn't need to go to the primary literature at all. I'll let them know!" Finding a needle in a haystack can be easy compared to the task of locating a particular number in the scientific literature. Our librarian wanted to know whether I had a starting point for her in the hunt. In this case, the solution was easy, but not every search will be so quick! The following brief introduction to searching chemical literature is directed specifically at physical chemistry. As you begin to do original research, these skills will need to be expanded and refined!

First of all, what is the chemical literature? It is all the journals, books, reviews, data compilations, and indices related to chemistry — and it is immense! More than 800 scientific journals are published every year, along with hundreds of books. One way of organizing the literature is hierarchical. At the base is the *primary literature:* original reports of research results written by the researchers themselves. Most often, these are in the form of journal articles, but they can also be monographs (books written by a single author or a small group of authors) or part of a symposium collection (a written collection covering talks given at a scientific meeting and edited by someone in the field). Drawing from the primary literature, a second tier of *reviews and compilations* is created. For example, every year a volume reviewing recent developments in computational chemistry is published. The National Institute for Science and Technology (NIST) regularly publishes compilations of data, such as tables of heats of formation, from the primary literature. Finally, at the top of the heap are *guides to the literature.* These are indices that can help you locate something in the primary or secondary literature. *Chemical Abstracts* has been indexing the chemical literature for a century! Computers have made using these indices much less arduous.

The first step in a successful literature search is to know what you are looking for. This seems obvious, but the more clearly you can define what you are looking for, the easier it will be to find. Are you looking for a method? for a particular value? for background information? It also helps to know whether there are alternative terms for your target. Knowing that methyl alcohol and methanol are the same thing will keep you from missing potentially useful references. Narrowing the subfield can also narrow the search. For example, knowing the methods by which bond lengths are typically measured can significantly limit your search for the HCl bond length.

Next, check for any leading references. Is a paper referenced in the lab handout? Are any names associated with a technique? Is any one person referred to? Given a reference to the literature, you can find related papers published both before and after it by using different indices.

Once you have a defined a search target and have collected any leading references, it's time to head for the library. Depending on what you are looking for, you might start in different places. If you are looking for background material in an area, you might begin with the collection catalog. A student in my graduate seminar, who was doing a project on QSAR parameters for sweetness, required some background on molecule theories of sweetness. The catalog steered her to several books on sweetness. Data compilations are a good first choice for finding individual values such as heats of formation, equilibrium constants, and average bond lengths. Good starting places are the CRC and NIST tables, which are now available on the Web (see Section 4–4).

Finally, a guide to the haystack is indispensable. Nothing can replace a good science librarian. She can help you make the best use of indices and databases and can help you refine and focus your search.

4–4 Resources

Good Writing Manuals

The ACS Style Guide: A Manual for Authors and Editors. Janet S. Dodd, editor. Washington, DC: American Chemical Society, 1997. This is a good basic reference that is probably in your library. It has a section on grammar that is short, but to the point, along with some very specific advice for a writing chemist.

Successful Lab Reports: A Manual for Science Students by Christopher S. Lobban and Maria Schefter. New York: Cambridge University Press, 1992.

The Craft of Scientific Writing by Michael Alley. New York: Springer-Verlag, 1996. This is a very thorough look at how to write in science — an excellent reference book for anyone planning to go on in science, either in graduate school or work.

Sources of Data

CRC Handbook of Chemistry and Physics. Cleveland, Ohio: CRC Press. This is your first stop for bits and pieces of chemical information — everything from the pH values of foods to nuclear data.

NIST Webbook is at http://webbook.nist.gov/. Print versions of the data in the *NIST Webbook* can be found in the *JANAF Thermochemical Tables* published by the American

Chemical Society and the American Institute of Physics for the National Bureau of Standards. This source is great for information on small molecules, such as I_2 and HCl.

Molecular Spectra and Molecular Structure. New York: Prentice-Hall, 1939–66. *Atomic Spectra and Atomic Structure,* by Gerhard Herzberg. New York: Dover Publications, 1944. These works contain data useful for many physical chemistry labs.

Quanta: A Handbook of Concepts by Peter Atkins. Oxford: Clarendon Press, 1974. This is an excellent and concise source of background information for quantum mechanics and spectroscopy.

Solutions to Example Problems in Chapter 2

· ·

Example 2.1

$$\sum_J (2J+1)e^{-\beta hcBJ(J+1)} = \sum_J 2Je^{-\beta hcBJ(J+1)} + \sum_J e^{-\beta hcBJ(J+1)}$$

$$= 2\sum_J e^{-\beta hcBJ(J+1)} + \sum_J e^{-\beta hcBJ(J+1)}$$

· ·

Example 2.2

Start by assuming that the summations begin at 1 and that each has N terms. These weren't specified, and you don't have a context to tell you otherwise. Now you have a more explicit version of the summation given.

$$\sum_{i=1}^{N}\sum_{j=1}^{N}\left(\int \psi_i^* \psi_j d\tau\right)$$

A good way to work with these sorts of sums is to lay out a matrix of terms, such as

$$
\begin{array}{cccc}
\int \psi_1^* \psi_1 d\tau & \int \psi_1^* \psi_2 d\tau & \int \psi_1^* \psi_3 d\tau & \cdots \\
\int \psi_2^* \psi_1 d\tau & \int \psi_2^* \psi_2 d\tau & \int \psi_2^* \psi_3 d\tau & \\
\int \psi_3^* \psi_1 d\tau & \int \psi_3^* \psi_2 d\tau & \int \psi_3^* \psi_3 d\tau & \\
\vdots & & & \ddots
\end{array}
$$

Now substituting for each of the integrals yields

$$
\begin{array}{cccc}
1 & 0 & 0 & \cdots \\
0 & 1 & 0 & \\
0 & 0 & 1 & \\
\vdots & & & \ddots
\end{array}
$$

From this you can see that the only non-zero terms are those along the diagonal, those with $i = j$. The double sum thus collapses to a single sum.

$$\sum_{i=1}^{N}\int \psi_i^* \psi_j d\tau = \sum_{i=1}^{N} 1$$
$$= N$$

What would the answer be if the sums started at 0? $N + 1$.

Example 2.3

$$x! = 1 \cdot 2 \cdot 3 \cdot \ldots \cdot x$$

In product notation this is

$$\prod_{i}^{x} i$$

Note that 0! can't be evaluated using this definition. It is by definition just 1.

Example 2.4

Just by eye, you would expect the expression

$$\sum_{i} e^{-\beta \varepsilon_i}$$

to converge. As each ε_i gets bigger, the term

$$1\!\!\Big/ e^{\beta \varepsilon_i}$$

gets smaller, as long as ε_i is positive. The first 10 terms of the summation and the values of the summation truncated after each term are as follows:

Number of Terms	Value of ith Term	Value of Sum of i Terms
1	0.2987	0.2987
2	0.08935	0.3883
3	0.02671	0.4150
4	0.007937	0.4230
5	0.002386	0.4253
6	0.0007133	0.4261
7	0.00002132	0.4263
8	0.00006374	0.4264
9	0.00001905	0.4264
10	0.000005695	0.4264

It's clear that after 10 terms the series has converged to four significant figures. Consider, however, the case where ν is 3.18×10^{11}Hz. Now the results are significantly different.

Number of Terms	Value of Sum
5	4.30
10	7.63
50	71.56
100	18.92
250	19.03
500	19.03
1,000	19.03

Now it requires 250 terms to reach convergence to four significant figures. Truncation must be handled individually for each case!

. . . .

Example 2.5

The trick here is to realize that

$$e^{-\beta i} = \left(e^{\beta}\right)^{-i}$$

If you then let $x = e^{\beta}$ you can write

$$\sum_{i=0}^{\infty} e^{-i\beta}$$

as

$$\sum_{i=0}^{\infty} x^{-i}$$

Then use the identity

$$\sum_{i=0}^{\infty} x^{-i} = \frac{1}{1-x}$$

substituting in for x to yield

$$\frac{1}{1-e^{\beta}}$$

• •

Example 2.6

In this case I used Mathematica to do the fits, using the commands

```
data={{1, 79.6204}, {1.1, 30.6865}, {1.2, 7.21718},
    {1.3, 0.}, {1.4, 3.01763}, {1.5, 12.3474}};
Fit[data, {x^2, x, 1}, x]
```

for the quadratic fit and

```
Fit[data, {x^3, x^2, x, 1}, x]
```

for the cubic fit. You could also use a spreadsheet such as Excel or many scientific calculators. The resulting quadratic fit is

$$1{,}262 - 1{,}895x + 709.4x^2$$

where the coefficient on the second-order term is 1,895 and the minimum is found by setting the first derivative to zero and solving for x, the HCl distance. The predicted value is 1.336 Å. The resulting cubic fit is

$$3{,}254 - 6{,}782x + 4{,}661x^2 - 1{,}054x^3$$

where the coefficient on the second-order term is 4,661 and the minimum is found by setting the first derivative to zero and solving for x, the HCl distance. The predicted value is 1.305 Å.

• •

Example 2.7

Using the definition of the Taylor series above, the expansion (to second order) is

$$f(x) \approx \frac{e^{-x^2}}{0!}\bigg|_{x=2} + \frac{(x-2)}{1!}(-2xe^{-x^2})\bigg|_{x=2} + \frac{(x-2)^2}{2!}(4x^2e^{-x^2} - 2e^{-x^2})\bigg|_{x=2}$$

$$= \frac{1}{e^4} - \frac{4}{e^4}(x-2) + \frac{7}{e^4}(x-2)^2$$

To fourth order, the approximation to the function looks like

$$f(x) \approx \frac{e^{-x^2}}{0!} + \frac{(x-2)}{1!}(-2xe^{-x^2}) + \frac{(x-2)^2}{2!}(4x^2e^{-x^2} - 2e^{-x^2}) +$$

$$\frac{(x-2)^3}{3!}(12xe^{-x^2} - 8x^3e^{-x^2}) + \frac{(x-2)^4}{4!}(12e^{-x^2} - 48x^2e^{-x^2} + 16x^4e^{-x^2})$$

$$= \frac{1}{e^4} - \frac{4}{e^4}(x-2) + \frac{7}{e^4}(x-2)^2 - \frac{20}{3e^4}(x-2)^3 + \frac{19}{6e^4}(x-2)^4$$

Remember to check the rules on differentiation if you have difficulty obtaining these terms. Recall also that 0! is 1.

I can get Mathematica to do much of this work for me. For example,

```
f[x_] := E^-x^2
f[2]/0! + ((x -2)((D[E^-x^2, x] /. x -> 2))) +
    (((x -2)^2)(D[f[x],{x, 2}] /.x -> 2)/2!) +
    (((x - 2)^3)(D[f[x], {x,3}] /. x -> 2)/ 3!) +
    (((x - 2)^4)(D[f[x], {x, 4}] /. x -> 2)/4!)
Plot[%, {x, 1, 3}]
```

will make a plot of the expansion to fourth order from $x = 1$ to $x = 3$

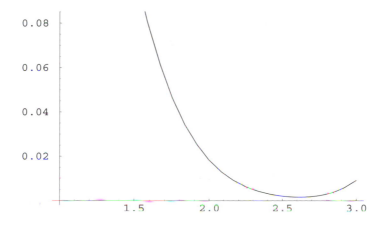

Comparing the expansion to second order with the actual function,

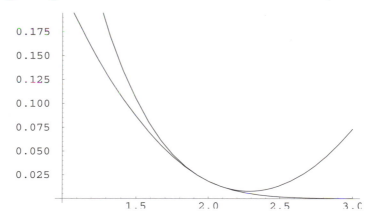

you can see that the function is well reproduced only between about 1.75 and 2.1. Using the longer expansion gives a much better description over a wider range.

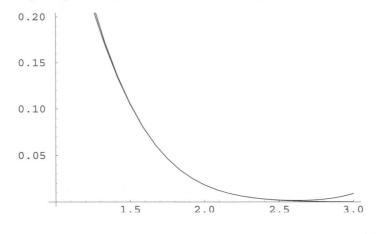

Example 2.8

Begin with the expression for a Fourier series.

$$f(x) = \sum_{n=0}^{\infty} a_n \cos(nx) + \sum_{n=0}^{\infty} b_n \sin(nx)$$

Because a_n is 0, this quickly simplifies to

$$f(x) = \sum_{n=0}^{\infty} b_n \sin(nx)$$

Only the odd-numbered terms survive in this sum (all the even b_n are zero), so change the summation to

$$f(x) = \sum_{j=0}^{\infty} \frac{4}{(2j+1)\pi} \sin((2j+1)x)$$

to account for this. Note that j still runs {0, 1, 2, 3, ...}, but this gives n values of $(2j + 1)$ or {1, 3, 5, ...}.

Here is the plot of the first five terms of the Fourier series:

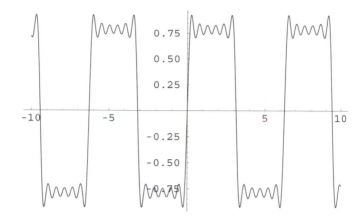

The first 20 terms give you

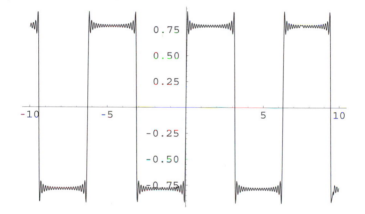

The first 100 terms actually give a pretty good representation:

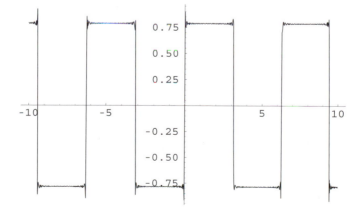

Imagine how well you could reproduce a "function" that didn't have corners!

· ·

Example 2.9

a. This is straightforward, if a bit tedious.

$$\frac{-\hbar^2}{2m}\left(\frac{\partial^2}{\partial x^2}+\frac{\partial^2}{\partial y^2}\right)\left[\frac{2}{L}\sin\left(\frac{n_1\pi x}{L}\right)\sin\left(\frac{n_2\pi y}{L}\right)\right]$$

$$=\frac{-\hbar^2}{2m}\left(\frac{\partial^2}{\partial x^2}\left[\frac{2}{L}\sin\left(\frac{n_1\pi x}{L}\right)\sin\left(\frac{n_2\pi y}{L}\right)\right]+\frac{\partial^2}{\partial y^2}\left[\frac{2}{L}\sin\left(\frac{n_1\pi x}{L}\right)\sin\left(\frac{n_2\pi y}{L}\right)\right]\right)$$

$$=\frac{-\hbar^2}{2m}\left[\frac{2}{L}\sin\left(\frac{n_2\pi y}{L}\right)\frac{\partial^2}{\partial x^2}\sin\left(\frac{n_1\pi x}{L}\right)+\frac{2}{L}\sin\left(\frac{n_1\pi x}{L}\right)\frac{\partial^2}{\partial y^2}\sin\left(\frac{n_2\pi y}{L}\right)\right]$$

$$=\frac{-\hbar^2}{2m}\frac{2}{L}\left[\sin\left(\frac{n_2\pi y}{L}\right)\frac{\partial}{\partial x}\frac{n_1\pi}{L}\cos\left(\frac{n_1\pi x}{L}\right)+\sin\left(\frac{n_1\pi x}{L}\right)\frac{\partial}{\partial y}\frac{n_2\pi}{L}\cos\left(\frac{n_2\pi y}{L}\right)\right]$$

$$=\frac{-\hbar^2}{2m}\frac{2}{L}\left[\sin\left(\frac{n_2\pi y}{L}\right)\frac{n_1\pi}{L}\left(-\frac{n_1\pi}{L}\sin\left(\frac{n_1\pi x}{L}\right)\right)+\sin\left(\frac{n_1\pi x}{L}\right)\frac{n_2\pi}{L}\left(-\frac{n_2\pi}{L}\sin\left(\frac{n_2\pi x}{L}\right)\right)\right]$$

$$=\frac{\hbar^2}{2m}\frac{2}{L}\left[\left(\frac{n_1\pi}{L}\right)^2+\left(\frac{n_2\pi}{L}\right)^2\right]\sin\left(\frac{n_1\pi x}{L}\right)\sin\left(\frac{n_2\pi y}{L}\right)$$

b. This requires using the chain rule. Break down

$$\frac{\partial}{\partial\beta}\left(\frac{1}{1-e^{-\beta\varepsilon}}\right)$$

as follows:

$$-\beta\varepsilon\longrightarrow1-e^{(\,)}\longrightarrow\frac{1}{(\,)}$$

Differentiate the "steps":

$$-\varepsilon\longrightarrow-e^{(\,)}\longrightarrow-\frac{1}{(\,)^2}$$

and multiply:

$$(-\varepsilon)\left(-e^{(-\beta\varepsilon)}\right)\left(-\frac{1}{(1-e^{-\beta\varepsilon})^2}\right)=-\frac{\varepsilon e^{(-\beta\varepsilon)}}{(1-e^{-\beta\varepsilon})^2}$$

c. This requires using the chain rule. Break down

$$\frac{\partial}{\partial \beta}\left(\ln \frac{1}{1-e^{-\beta\varepsilon}}\right)$$

as follows:

$$-\beta\varepsilon \longrightarrow 1-e^{(\ )} \longrightarrow \frac{1}{(\ )} \longrightarrow \ln(\ )$$

Differentiate the "steps":

$$-\varepsilon \longrightarrow -e^{(\ )} \longrightarrow -\frac{1}{(\ )^2} \longrightarrow \frac{1}{(\ )}$$

and multiply:

$$(-\varepsilon)\left(-e^{(-\beta\varepsilon)}\right)\left(-\frac{1}{(1-e^{-\beta\varepsilon})^2}\right)\frac{1}{\left(\frac{1}{1-e^{-\beta\varepsilon}}\right)} = -\frac{\varepsilon e^{(-\beta\varepsilon)}}{(1-e^{-\beta\varepsilon})}$$

· ·

Example 2.10

To find any critical point (including the minimum) take the first derivative and set it equal to zero, so

$$\frac{\partial}{\partial r}R_{2s}(r) = \frac{\partial}{\partial r}\left\{\frac{1}{\sqrt{8}}\left(\frac{Z}{a_o}\right)^{3/2}\left(2-\frac{Zr}{a_o}\right)e^{-Zr/2a_o}\right\}$$

Pull out the constants and differentiate. You need to use the product rule here.

$$=\frac{1}{\sqrt{8}}\left(\frac{Z}{a_o}\right)^{3/2}\frac{\partial}{\partial r}\left\{\left(2-\frac{Zr}{a_o}\right)e^{-Zr/2a_o}\right\}$$

$$=\frac{1}{\sqrt{8}}\left(\frac{Z}{a_o}\right)^{3/2}\left\{\left(2-\frac{Zr}{a_o}\right)\frac{\partial}{\partial r}e^{-Zr/2a_o} + e^{-Zr/2a_o}\frac{\partial}{\partial r}\left(2-\frac{Zr}{a_o}\right)\right\}$$

$$=\frac{1}{\sqrt{8}}\left(\frac{Z}{a_o}\right)^{3/2}\left\{\left(2-\frac{Zr}{a_o}\right)\left(-\frac{Z}{2a_o}e^{-Zr/2a_o}\right) + e^{-Zr/2a_o}\left(-\frac{Z}{a_o}\right)\right\}$$

$$=-\frac{1}{\sqrt{8}}\left(\frac{Z}{a_o}\right)^{5/2}\left(2-\frac{Zr}{2a_o}\right)e^{-Zr/2a_o}$$

Set this equal to zero and solve.

$$0 = -\frac{1}{\sqrt{8}}\left(\frac{Z}{a_o}\right)^{5/2}\left(2 - \frac{Zr}{2a_o}\right)e^{-Zr/2a_o}$$

$$\frac{4a_o}{Z} = r$$

To confirm that this really is a minimum, you have to take the second derivative and show that it is positive at $r = 4a_0/Z$. (Most of the time you won't bother with this.)

$$\frac{\partial^2}{\partial r^2}\left\{\frac{1}{\sqrt{8}}\left(\frac{Z}{a_o}\right)^{3/2}\left(2 - \frac{Zr}{a_o}\right)e^{-Zr/2a_o}\right\}\Bigg|_{r=\frac{4a_o}{Z}} = \frac{Z^2\left(\frac{Z}{a_o}\right)^{3/2}}{\sqrt{8}a_o^2 e^2}$$

Because Z, a_0, and e are all positive, this is also positive, confirming that the critical point is a minimum.

· ·

Example 2.11

a.

$$\langle r\rangle_{1s} = \frac{1}{\pi a_o^3}\int_0^\infty\int_0^\pi\int_0^{2\pi} r^3 e^{-2r/a_o}\sin\theta\, d\phi\, d\theta\, dr$$

This can be broken up into a product of integrals:

$$\frac{1}{\pi a_o^3}\left\{\int_0^\infty r^3 e^{-2r/a_o}dr \times \int_0^\pi \sin\theta\, d\theta \times \int_0^{2\pi}d\phi\right\}$$

The last two are trivial to evaluate.

$$\frac{1}{\pi a_o^3}\left\{\int_0^\infty r^3 e^{-2r/a_o}dr \times 2 \times 2\pi\right\}$$

The first one can be solved using a table of integrals and a simple substitution.

$$\int_0^\infty r^3 e^{-2r/a_o}dr = \int_0^\infty\left(\frac{a_o}{2}u\right)^3 e^{-u}\left(\frac{a_o}{2}du\right) = \left(\frac{a_o}{2}\right)^4\int_0^\infty u^3 e^{-u}du$$

A table of integrals gives

$$\int_0^\infty x^n e^{-ax} dx = \frac{n!}{a^{n+1}}$$

for $a > 0$ and n a positive integer. For $n = 3$ and $a = 1$,

$$\int_0^\infty u^3 e^{-u} du = 3! = 6$$

Therefore, the final integral is

$$\frac{1}{\pi a_o^3} \left\{ \left(\frac{a_o}{Z}\right)^4 \times 6 \times 2 \times 2\pi \right\} = \frac{3}{2} a_o$$

. . . .

Example 2.12

The key is always the selection of u and dv. Sometimes the integral can be divided in more than one way. If this is the case, and you find that the way you chose isn't working out cleanly, try again with another split of the integral.

Here one possible choice is

$$u = x$$
$$dv = e^{-2x} dx$$

This leads to

$$du = dx$$
$$v = \int e^{-2x} dx = -\frac{1}{2} e^{-2x}$$

The integration by parts is thus

$$\int u\, dv = uv - \int v\, du = (x)\left(-\frac{1}{2}e^{-2x}\right) - \int \left(-\frac{1}{2}e^{-2x}\right) dx$$

Integrating yields

$$(x)\left(-\frac{1}{2}e^{-2x}\right) - \left(-\frac{1}{2}\right)\left(-\frac{1}{2}e^{-2x}\right) = -\frac{1}{2}e^{-2x}\left(x + \frac{1}{2}\right)$$

. . . .

Example 2.13

Rearrange the equation to get t on one side and [B] on the other

$$\frac{d[B]}{k_1\{[A]_o - [B]\} - k_2[B]} = dt$$

Now integrate

$$\int_0^{B_t} \frac{d[B]}{k_1\{[A]_o - [B]\} - k_2[B]} = \int_o^t dt$$

$$\int_0^{B_t} \frac{d[B]}{k_1[A]_o - (k_1 + k_2)[B]} = \int_o^t dt$$

$$\frac{\ln(-k_1[A]_o) - \ln(-k_1[A]_o + (k_1 + k_2)B_t)}{(k_1 + k_2)} = t$$

$$\ln\left(\frac{k_1[A]_o}{k_1[A]_o - (k_1 + k_2)B_t}\right) = (k_1 + k_2)t$$

Now solve for B_t

$$\frac{k_1[A]_o}{k_1[A]_o - (k_1 + k_2)B_t} = e^{(k_1+k_2)t}$$

$$k_1[A]_o = e^{(k_1+k_2)t}(k_1[A]_o - (k_1 + k_2)B_t$$

$$k_1[A]_o = e^{(k_1+k_2)t}k_1[A]_o - e^{(k_1+k_2)t}(k_1 + k_2)B_t$$

$$\frac{e^{(k_1+k_2)t}k_1[A]_o - k_1[A]_o}{e^{(k_1+k_2)t}(k_1 + k_2)} = B_t$$

. . . .

Example 2.14

$$\frac{-\hbar^2}{2m}\frac{d^2\psi(x)}{dx^2} = E\psi(x)$$

$$\frac{-\hbar^2}{2m}\frac{d^2\psi(x)}{dx^2} + 0\cdot\frac{d^2\psi(x)}{dx^2} - E\psi(x) = 0$$

The auxiliary equation is just

$$\frac{-\hbar^2}{2m}s^2 + 0 \cdot s - E = 0$$

$$s^2 + \left(\frac{2m}{\hbar^2}\right)E = 0$$

The roots of the auxiliary equation are

$$s_1 = \frac{(2mE)^{\frac{1}{2}}}{\hbar}i \qquad\qquad s_2 = -\frac{(2mE)^{\frac{1}{2}}}{\hbar}i$$

so the solution is

$$\psi(x) = c_1 e^{\frac{(2mE)^{\frac{1}{2}}}{\hbar}ix} + c_2 e^{\frac{-(2mE)^{\frac{1}{2}}}{\hbar}ix}$$

The coefficients c_1 and c_2 are determined by using the boundary conditions for the problem.

••

Example 2.15

Starting with

$$\frac{-\hbar^2}{2m}\frac{d^2\psi(x)}{dx^2} + bx = E\psi(x)$$

substitute in for $\psi(x)$ and its second derivative as follows

$$\psi(x) = a_o + a_1 x + a_2 x^2 + \cdots = \sum_{i=0}^{\infty} a_i x^i$$

$$\frac{d^2\psi(x)}{dx^2} = 2a_2 + 6a_3 x + 12a_4 x^2 \cdots = \sum_{i=2}^{\infty} i(i-1)a_i x^{i-2}$$

which yields

$$\frac{-\hbar^2}{2m}\sum_{i=2}^{\infty} i(i-1)a_i x^{i-2} + bx - E\sum_{i=0}^{\infty} a_i x^i = 0$$

Change to the same index range (0 to ∞) so that the two sums can be combined. The trick is to substitute $k = i - 2$

$$\sum_{k=0}^{\infty}\frac{-\hbar^2}{2m}(k+1)(k+2)a_{k+2}x^k + bx - E\sum_{i=0}^{\infty} a_i x^i = 0$$

Since the index in the second summation can just be changed to k,

$$\sum_{k=0}^{\infty} \frac{-\hbar^2}{2m}(k+1)(k+2)a_{k+2}x^k + bx - E\sum_{k=0}^{\infty}a_k x^k = 0$$

the sums can easily be combined:

$$\left[\sum_{k=0}^{\infty}\frac{-\hbar^2}{2m}(k+1)(k+2)a_{k+2}x^k - Ea_k x^k\right] + bx = 0$$

Now the solutions to this are those where the coefficients of x^k are 0. The tricky part here is that the coefficient on x^1 comes from both the summation and b! So take the first two terms separately.

For $k = 0$,

$$-\frac{\hbar^2}{2m}(1)(2)a_2 - Ea_o = 0$$

$$a_2 = \frac{-mE}{\hbar^2}a_o$$

For $k = 1$,

$$-\frac{\hbar^2}{2m}(2)(3)a_3 - Ea_1 + b = 0$$

$$a_3 = \frac{2m(b - Ea_o)}{3\hbar^2}$$

The remaining terms ($k > 1$) are

$$\frac{-\hbar^2}{2m}(k+1)(k+2)a_{k+2} - Ea_k = 0$$

$$a_{k+2} = E\left(\frac{-2m}{\hbar^2}\right)\left(\frac{1}{(k+1)(k+2)}\right)a_k$$

A more convenient form for a_{k+2} is obtained by setting $j = k + 2$

$$a_j = -E\left(\frac{2m}{\hbar^2}\right)\left(\frac{1}{j(j-1)}\right)a_{j-2}$$

The final form for the wavefunction would thus be

$$\psi(x) = a_o + a_1 x - \frac{mE}{\hbar^2}a_o x^2 + \frac{2m(b - Ea_1)}{3\hbar^2} - \frac{2mE}{\hbar^2}\sum_{j=4}^{\infty}\frac{1}{j(j-1)}a_{j-2}x^j$$

Example 2.16

Using

$$x_{n+1} = x_n + h$$
$$f(x_{n+1}) = f(x_n) + hZ(x_n, f(x_n))$$

the following table is helpful.

Time	Numerical Solution	Analytical Solution
$t = 0$	$B(0) = 0$	
$t = 0.1$	$B(0.5) = B(0) + 0.1 \bullet \{k_1[A_0 - B(0)] - k_2B(0)\}$	1.10
	$= 0 + (0.1)\{2(7 - 0) - 3(0)\}$	
	$= 1.4$	
$t = 0.2$	2.1	1.77
$t = 0.3$	2.45	2.18
$t = 0.4$	2.625	2.42
$t = 0.5$	2.7125	2.57

The approximation isn't terribly good. If you choose a smaller time step, 0.01, the results are much better.

Time	Numerical Solution	Analytical Solution
0.01	0.14	0.14
0.02	0.27	0.27
0.03	0.40	0.39
0.04	0.52	0.51
0.05	0.63	0.62

If the time step is too large, such as 0.5, the answers become nonsensical, with the amount of B produced exceeding the amount of A you began with! These approaches must be used with care.

Example 2.17

The three students are equivalent to three "boxes" in which five indistinguishable objects (one brown M&M is the same as another) will be arranged. The number of possible ways in which this can be done is

$$\frac{(M + N - 1)!}{(M - 1)!N!}$$

where $M = 3$ and $N = 5$

$$\frac{(3+5-1)!}{(3-1)!5!} = 21$$

If the M&Ms are all of different colors, they can be distinguished one from another, so the number of possible arrangements increases. For example, if student A had 3 brown, student B had 2 brown, and student C didn't have any, that's one possible arrangement. If the M&Ms were different colors, then student A could have a blue, a brown, and red *or* a yellow, a green, and a blue *or* a red, a blue, and a yellow, etc. The number of arrangements is given by

$$M^N = 3^5 = 243$$

· ·

Example 2.18

To find $S+$:

$$S_+ = S_x + iS_y$$

$$= \begin{pmatrix} 0 & \frac{1}{2}\hbar \\ \frac{1}{2}\hbar & 0 \end{pmatrix} + i\begin{pmatrix} 0 & -\frac{1}{2}i\hbar \\ \frac{1}{2}i\hbar & 0 \end{pmatrix}$$

$$= \begin{pmatrix} 0 + 0i & \frac{1}{2}\hbar - \frac{1}{2}i^2\hbar \\ \frac{1}{2}\hbar + \frac{1}{2}i^2\hbar & 0 + 0i \end{pmatrix}$$

$$= \begin{pmatrix} 0 & \hbar \\ 0 & 0 \end{pmatrix}$$

Matrix multiply to find S_xS_y:

$$S_xS_y = \begin{pmatrix} 0 & \frac{1}{2}\hbar \\ \frac{1}{2}\hbar & 0 \end{pmatrix}\begin{pmatrix} 0 & -\frac{1}{2}i\hbar \\ \frac{1}{2}i\hbar & 0 \end{pmatrix}$$

$$= \begin{pmatrix} 0 \cdot 0 + \frac{1}{2}\hbar \cdot \frac{1}{2}\hbar & 0 \cdot -\frac{1}{2}i\hbar + \frac{1}{2}\hbar \cdot 0 \\ \frac{1}{2}\hbar \cdot 0 + 0 \cdot \frac{1}{2}\hbar & \frac{1}{2}\hbar \cdot -\frac{1}{2}i\hbar + 0 \cdot 0 \end{pmatrix}$$

$$= \begin{pmatrix} \frac{1}{4}i\hbar^2 & 0 \\ 0 & -\frac{1}{4}i\hbar^2 \end{pmatrix}$$

To show that S_y is self-adjoint:

$$\left(S_y^t\right)^* = \left[\begin{pmatrix} 0 & -\tfrac{1}{2}i\hbar \\ \tfrac{1}{2}i\hbar & 0 \end{pmatrix}^t\right]^*$$

$$= \left[\begin{pmatrix} 0 & \tfrac{1}{2}i\hbar \\ -\tfrac{1}{2}i\hbar & 0 \end{pmatrix}\right]^*$$

$$= \begin{pmatrix} (0)^* & \left(\tfrac{1}{2}i\hbar\right)^* \\ \left(-\tfrac{1}{2}i\hbar\right)^* & (0)^* \end{pmatrix}$$

$$= \begin{pmatrix} 0 & -\tfrac{1}{2}i\hbar \\ \tfrac{1}{2}i\hbar & 0 \end{pmatrix}$$

$$= S_y$$

. . . .

Example 2.19

Plotting the entire set of data given and doing a least-squares fit to a line yields a calibration equation of the form T = 1.03V − 0.018. A plot of this line superimposed on the data, along with a cursory examination, suggests that a better line might be obtained if the first two data points were dropped.

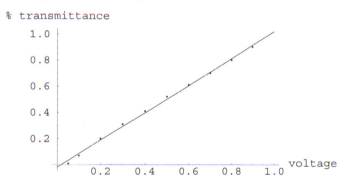

In this case, the resulting linear calibration equation would be T = 0.99 V + 0.012. Note that the fit to the majority of points in the middle of the calibration range is visibly improved in this case.

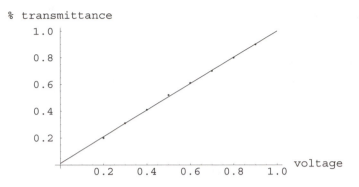

If one were working roughly in the range of 20% to 90% transmittance, the second calibration curve would provide better results, whereas if one were working over a wider range of transmittances, the first relationship would be more appropriate. Note that it was important to plot the data! Resist the temptation to skip this step. It can reveal useful information (in this case suggesting the elimination of the low-voltage points) as well as help you catch errors in data input. Obviously, the final choice of equation depends on the experiments planned and will affect the accuracy of your results.

Example 2.20

Answers will vary.

Example 2.21

Note that all three plots appear to be linear. The value of the rate constant calculated for data set 1 (2.33×10^{-4}/sec), however, differs substantially from those calculated for the other two data sets (7.25×10^{-4}/sec and 7.21×10^{-4}/sec, respectively). Because the value of the rate constant should depend only on the temperature, not on the initial pressures, this is surprising. One might suspect that some kind of undetected blunder was made during the first experiment, but it turns out that the discrepancy is due to the failure of the model used. The kinetics of cyclopropane isomerization are first order at high pressures, but at lower pressures they are second order. Over the limited time interval used in this experiment, the data appear to be linear.

From data set 1:

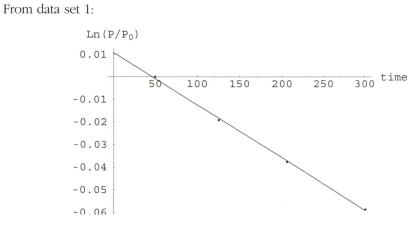

From data set 2:

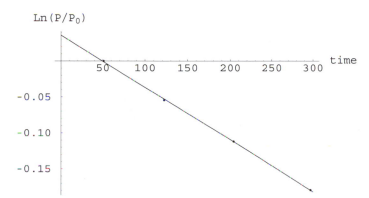

From data set 3:

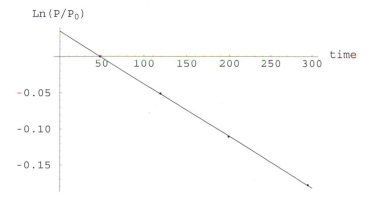

Example 2.22

5,039 kJ/mol. Watch out for significant figures; don't report this as 5,039.2!

Example 2.23

294 kJ/mol.

Example 2.24

627 kJ/mol. You would report the value as 5,039 ± 627 kJ/mol.

Example 2.25

The value that seems most likely to be an outlier is 6,076. The mean without the outlier is 4,780 kJ/mol. The average deviation from the new mean is 80 kJ/mol. The potential outlier is 1,296 kJ/mol away from the new mean. That is greater than 4×80 kJ/mol, so this point may be discarded.

Example 2.26

4,780 ± 167 kJ/mol. This value is more precise than the original, although, interestingly, it is further from the literature values, which makes it less accurate.

Example 2.27

The error in the slope of the first data set is 4×10^{-6}/sec; that of the second is 2×10^{-6}/sec. Note that these numbers don't tell you anything about how good a fit to a *line* you have. They only report how much the error in the dependent variable — in this case $\ln(P/P_0)$ — influenced the error in the slope. Because the error in the dependent variable for the first data set is larger than that for the second, the resulting error in the slope for the first set is also larger.

Example 2.28

0.07°C. They are all the same!

Example 2.29

To the one significant figure required, the values are all the same in this case (0.0007 torr). This may not always be true, so it should be calculated for all values.

Example 2.30

The expression to be used is:

$$\sqrt{\frac{\varepsilon_{enol}R^2T^2}{[\text{enol}]^2} + \frac{\varepsilon_{keto}R^2T^2}{[\text{keto}]^2} + \varepsilon_T R^2 l\left(n\frac{[\text{enol}]}{[\text{keto}]}\right)^2}$$

where the ε are the errors in the concentrations and in T. The final value is 7.84 ± 0.04 kJ/mol. Watch significant figures here as well!

Index